Acoustic Guitar Design

Richard Mark French

Acoustic Guitar Design

Springer

Richard Mark French
School of Engineering Technology
Purdue University West Lafayette
West Lafayette, IN, USA

ISBN 978-3-030-89380-4 ISBN 978-3-030-89381-1 (eBook)
https://doi.org/10.1007/978-3-030-89381-1

This Springer imprint is published by the registered company Springer Nature Switzerland AG
The registered company address is: Gewerbestrasse 11, 6330 Cham, Switzerland

As I look back, my time as a luthier divides itself clearly into the period before I was part of the Guild of American Luthiers and that afterward. With continuing thanks, I dedicate this book to the Guild and its members.

Foreword

I've never met an oboe maker. Most people live their lives not knowing how these complex and demanding instruments are played, much less how they're made.

When I started my career in guitars in the early 1970s, you could pretty much say the same about guitar makers. There were plenty of guitars available, but there wasn't a wide discussion possible about how—or why—they were made the way they were.

It was also pretty rare to meet a guitar maker, and big companies certainly weren't posting factory tours on YouTube. If you wanted to know how or why your guitar worked or if you actually wanted to make one, you were off on a Quest.

Now, of course, there are dozens of guitar construction books available and there's no shortage of opinions about why the originals worked (or didn't) or how new versions can or should be built.

Classical guitars, steel string instruments, and arch tops may all be acoustics, but their designs and construction utilize different philosophies and techniques. Until now, there hasn't been a book that explains and compares them.

My friend Mark French brings a unique and valuable set of talents to this little party. He's an accomplished engineer and an entertaining teacher who's taught hundreds of us how to use material science and math and apply them to musical instruments. I've sat in many of Mark's classes, and I never fail to learn from him.

In Acoustic Guitar Design, Mark shows us in clear text how these different guitars are designed and built, but he also teaches us the science behind these choices. It's a valuable addition to a huge subject, and well worth your time!

Tim Shaw
Chief Engineer – Guitars
Fender Musical Instruments
Nashville, TN, USA

Acknowledgments

A book like this is only possible with the contributions of many people who were willing to support an author or to serve the larger community of luthiers. I hardly know where to start in acknowledging the help I've received in writing this book and any list is necessarily incomplete. I hope those deserving people whose names don't appear here can forgive the omission.

Tim Shaw is an accomplished designer and a central figure in the guitar world. His knowledge of guitars and the guitar industry is extensive, and he's been generous in sharing it with me. He was an early supporter of the guitar lab at Purdue, benefitting many students.

Bob Taylor seems to be a force of nature. He co-founded Taylor Guitars and has done more than his part to advance guitar manufacturing technology. Along the way, he has freely shared his knowledge, even with his competitors. He has turned his attention to how guitar manufacturing can be more environmentally sustainable. More personally, he has generously supported the guitar lab at Purdue and contributed to both this and my previous books.

Andy Powers is very good guitar designer and one of the creative forces behind a new generation of Taylor guitars. He has generously shared his knowledge, even when we both knew he had other pressing tasks.

Richard Bruné is a builder and restorer of the highest order, and a respected historian of Spanish guitars. The lutherie world is surely on firmer ground than it was before he joined it. He routinely wrote detailed and thoughtful answers to even my simplest questions. I drew heavily on his expertise and he kindly reviewed the manuscript.

Doug Hunt is a valued colleague and friend who I first met as part of the STEM Guitar Project. He is a capable luthier as well as a talented musician. He talked with me for long, happy hours about some of the ideas presented here, helping me refine what I wanted to say.

R.M. Mottola has been a generous and patient source of information about the technical details of guitar making. I've asked him many questions and gotten thoughtful, informed answers. He diligently reviewed a draft of the book and made many useful suggestions.

Tim Olsen is the orbital center of the Guild of American Luthiers. I had to look up his actual title, which is founding editor, but that barely hints at the breadth of his job or his influence. He nurtures and cajoles recalcitrant luthiers into writing articles, and has brought American Lutherie from a newsletter to a respected professional journal with readers all over the world. He is also an ardent supporter of the Oxford comma.

Charles Fox is a top luthier. However, he is also a teacher and a designer of tools that help the rest of us make better guitars. He has taught me much. There are few better ways to spend an hour than talking guitars with Charles. I look forward to more of those.

Paco Chorobo is a Spanish builder of fine classical and flamenco guitars. I met him as a student in his flamenco guitar class, where I learned much. He takes his craft seriously while exuding a happy enthusiasm that wears off on his students.

Davin Huston is a colleague and friend in the School of Engineering Technology. He has helped me stay on an even keel while writing this book, sometimes listening to me fret about the lot of the author.

Ken Burbank is about the best boss a professor could ask for. In a large department within a national research university, he has supported me as I've worked on guitar design and manufacturing while the rest of his faculty did something normal. He cares deeply about his faculty and supports us, even when we occasionally give him reason not to.

May everyone know someone like Kay Solomon. She breathes in literature as one would oxygen and love of the written word permeates her life. Inexplicably, she also enjoys line editing manuscripts and has now helped me with four books.

The National Science Foundation has supported the STEM Guitar Project for 10 years now, providing several million dollars to a team that uses guitar making to teach technical subjects to high school students around the country. This project has helped me develop as a professor, a luthier, and an engineer.

The STEM Guitar Project Team, especially Tom Singer, Mike Aikens, Scot Rabe, Steve Brown, Doug Hunt, and Debbie French, are both colleagues and friends. Working with this group has been a high point in my career as a professor. I've learned much from them, some of which appears in this book.

I must thank the many people who have kindly shared pictures with me. Writing a book takes a long time and some of that time is spent finding pictures and securing permission to use them. In almost every instance, the community of luthiers, music stores, and manufacturers were generous and supportive.

My beloved wife, Amy, has now endured four books and no longer rolls her eyes audibly when I tell her I don't want to write another one.

Contents

Chapter 1
Introduction

The speed of sound in air is about 1100 ft./s (343 m/s).

The density of steel is about 0.28 lb/in.^{3} (7850 kg/m^3).

The density of spruce is around 0.015 lb/in.^{3} (415 kg/m^3).

The density of nylon is around 0.042 lb/in.^{3} (1150 kg/m^3).

Most people are between 5 and 6 ft. tall (152–183 cm).

The frequency of A_4 is defined as 440 Hz.

These basic facts dictate the size of acoustic guitars. These facts along with some history and our sense of aesthetics define the shape. These facts, along with some basic physics, determine much of the rest of the design of acoustic guitars. Figure 1.1 shows a particularly successful acoustic guitar, one that has affected many of the designs that came after it.

The acoustic guitar is a collection of solutions to opposing problems. The strings must have enough tension to be at the correct pitch and the structure of the instrument must be strong enough to bear these loads for decades, sometimes under hard use. At the same time, it must be light enough and flexible enough to respond to the vibration of the strings and make music.

Guitars are also a collection of approximations. The musical scale rendered into hardware by the spacing of the frets approximates an appealing, but impractical Pythagorean ideal. The calculated fret locations also rely on an approximate mathematical description of how strings vibrate.

Acoustic guitars also highlight how little we know about how to quantify sounds. There is not yet any robust way to calculate the sound quality of a guitar. While any competent player knows a good guitar when they play it, there is not yet any good mathematical description of what that means.

For all this, good luthiers routinely produce fine guitars, sometimes along traditional lines and sometimes by following their own creative paths. Even more, they sometimes perform this magic with only modest materials and tools.

The design of acoustic guitars has evolved over hundreds of years and that evolution has certainly accelerated. Still, the classical guitar was in its most familiar

R. M. French, *Acoustic Guitar Design*,
https://doi.org/10.1007/978-3-030-89381-1_1

Fig. 1.1 A 1942 Martin OOO-28 (image courtesy of Mike and Mike's Guitar Bar, mmguitarbar.com)

modern form by the late 1800s and the steel string acoustic guitar was in its familiar modern form before the middle of the twentieth century. While some builders are happily making fine copies of iconic instruments, others are modifying them or starting fresh with completely new designs.

Guitars have always been made of wood—usually a small number of species. Initially, builders used wood because it was the only practical material available. But, even now when other materials are available, almost all acoustic guitars are made of wood. The difference now is that some traditional species are endangered and protected by international agreement. In response, builders are turning to more sustainable species while some are exploring other materials.

Even though modern guitars usually follow a few familiar patterns, luthiers are an experimental bunch, generally unwilling to leave well enough alone just because a design works fine. Indeed, we may be in a golden age of guitar making as centuries of accumulated experience are now freely available to anyone who cares to learn it. This is combined with international supply chains that offer specialized tools and materials to anyone with web access and a modest budget. There are likely more luthiers now practicing that at any time in history. However, designing guitars and making them are different.

This is a guide for guitar makers, even new ones, who want to design their own instrument. I've tried to collect the important ideas and needed reference information in one place. There is little math and none beyond what you learned in Junior High School, so you can do all the calculations you need on the simplest of spreadsheets.

Once you've designed and made your own guitar, perhaps along conventional lines, you'll be ready to refine your design or to venture off in your own directions. You'll have the expertise you need to experiment with new designs. May you find your own artistic and musical voices in the instruments you've brought into being.

Chapter 2
Background

At its most basic, an acoustic guitar consists of a hollow body with a vibrating top (the soundboard), a neck with frets and a headstock where the tuners are mounted. Instruments that fit this description have existed for a very long time. Indeed, music seems a fundamental part of being human. Before we can start on the design process, it's important to learn something about where this combination of parts that form the guitar came from and how modern acoustic guitars acquired their familiar forms.

Archeologists speculate that the first stringed musical instruments may have been hunting bows fitted with some hollow object to act as a crude body. Musical bows are still in occasional use, often in Africa. These instruments usually use either some type of small gourd, as shown in Fig. 2.1, or the player's mouth as a resonant chamber. Whatever their exact forms, the first stringed musical instruments long predate the written word. Note that the instrument here follows traditional designs, but is a modern one, fitted with a geared bass tuner.

Stringed instruments are not just old, but widespread as well. Societies in many different places developed them and it isn't reasonable to think all these people were in contact with each other when they started. That means there isn't one single place where the first stringed instrument was made, just like there was no single place where writing started. Some ideas are so good, that they naturally occur to people separately. Stringed instruments appear to be one of those.

While stringed instruments may have arisen independently in different places, it didn't stay that way. Nomadic people likely spread their instruments as they traveled. Later, the same traders who brought silk and spices to Europe spread stringed instruments through Asia and Europe as well, and the result was broad families of stringed instruments preceding the guitar. Just as there is no single path from one side of a large forest to another, there is no single path from any one ancient stringed instrument to modern guitars. The story of the guitar has many threads, often interwoven. For example, I can't imagine trying to draw a single clear timeline from

R. M. French, *Acoustic Guitar Design*,
https://doi.org/10.1007/978-3-030-89381-1_2

Fig. 2.1 A musical bow (image courtesy of the Musical Instrument Museum, Phoenix, Arizona, USA, mim.org)

some ancient proto-instrument to the Taylor A12 that sits next to my desk as I write this.

Still, it helps to have some kind of historical framework in mind as we learn to design guitars, even if it is necessarily approximate. Perhaps it is enough to touch on a few important points along the historical paths leading to modern instruments, while remembering that whole books have been written about details I've skipped here. For now, we just need the overall trajectory.

Figure 2.2 sketches a simplified path that connects a few stepping stones on the way to modern guitars. It leads from the beginnings of stringed instruments to the three most common types of modern acoustic guitars: the nylon stringed classical, the steel stringed flattop and the archtop.

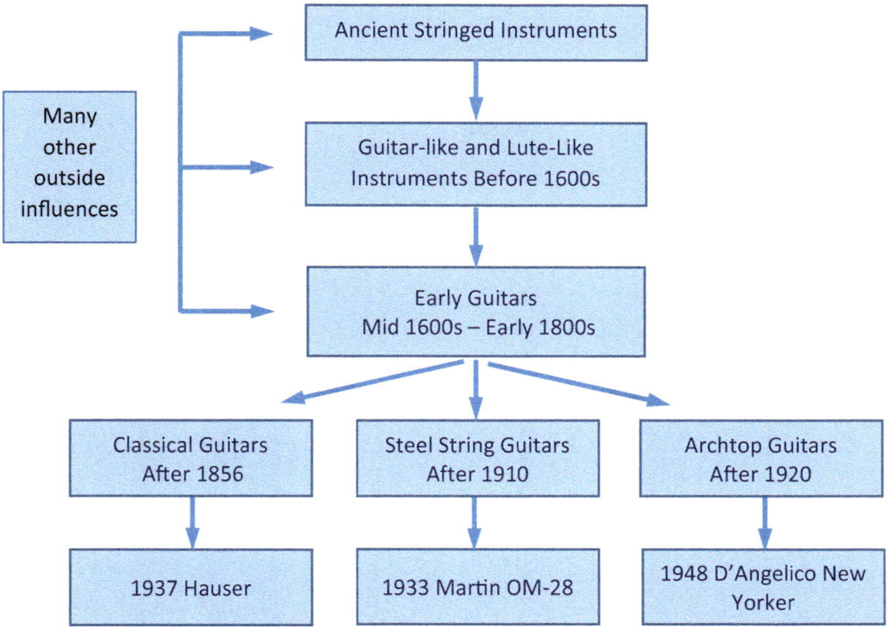

Fig. 2.2 A path toward modern instruments

Our timeline ends with three iconic guitars that we'll examine in detail later: the 1937 Hauser, the 1933 Martin OM-28 and the 1948 D'Angelico New Yorker. These instruments are well documented and have influenced newer designs. Guitar designers familiar with these three instruments should be able to move on to more modern designs and to innovate on their own. These are clearly not the only three instruments I could have chosen. Still, they are reasonable choices. While writing this book, I talked with skilled guitar designers and these three instruments were the result. My apologies if your favorite instrument is not among them.

2.1 Stringed Instruments Before the Guitar

People have played instruments that bear at least a rough resemblance to guitars for a very long time. Good designers must recognize that they are carrying on and perhaps adding to a tradition that has grown over millennia. The first stringed instruments with bodies, necks, and headstocks appeared more than 3000 years ago. A good example is the tanbur, that appears to date originally from about 1500 BCE. In spite of its ancient origins, the tanbur is still being played and a quick search shows a surprising number of builders making new ones. Figure 2.3 shows a Kurdish tanbur. It's definitely not a guitar, but has some of the necessary elements. The body has a flat soundboard, though with a rounded back. It has a long neck with tied

Fig. 2.3 A Kurdish Tanbur (Wikimedia Commons, commons.wikimedia.org, uploaded by user allauddin)

frets—they are not pieces of wire set into slots on the neck, as on a modern guitar, but are rather loops of heavy string wrapped around the neck at the right locations. The strings are tuned using wooden pegs that hold the needed tension by simple friction, akin to violin pegs.

The name guitar has origins almost as mixed as the instrument itself. The word guitar probably evolved from the Greek kithara (or cithara), an instrument in the lyre family. Instruments called guitarra (the guitarra latina and guitarra morisca) existed in Spain by 1200.

2.2 Early Guitars

One of the first instruments that was clearly called a guitar was the Baroque guitar, which was most popular from approximately 1600–1750. Figure 2.4 shows a Baroque guitar made by Mateo Seelos sometime before 1653. The fretboard on this instrument may not be original, as the top would likely have originally extended to around the ninth fret. The bridge may also not be original.

While different from modern instruments, it has features we recognize. The body has the now familiar form with an upper and lower bout, separated by a narrower waist. The single centered soundhole, just above the waist, is a feature on almost all modern guitars, as is the neck with angled headstock. The sides are made of bent wood and the top is made of softer wood from a conifer (evergreen tree).

We can't stretch the comparison too far because this instrument has features that almost never appear on modern guitars. The ten strings are in five pairs called courses. The soundhole rosette inlaid into the top around the soundhole accompanies a very elaborate decoration made of parchment that fills the soundhole. Also, the fretboard is level with the soundboard and additional frets are set directly into the top. Finally, the strings are tuned with wood friction pegs.

There are differences on the inside as well. The tops of baroque guitars are almost always reinforced with parallel wood bars perpendicular to the grain. This is called ladder bracing, presumably because the braces resemble the rungs of a ladder.

Fig. 2.4 A Baroque Guitar Made by Mateo Seelos before 1653 (Wikimedia Commons, commons.wikimedia.org, original photo by Olav Nyhus)

Fig. 2.5 The braced soundboard of a modern reproduction baroque guitar (Photo by Emily Shaw, emilyshawguitar.ca)

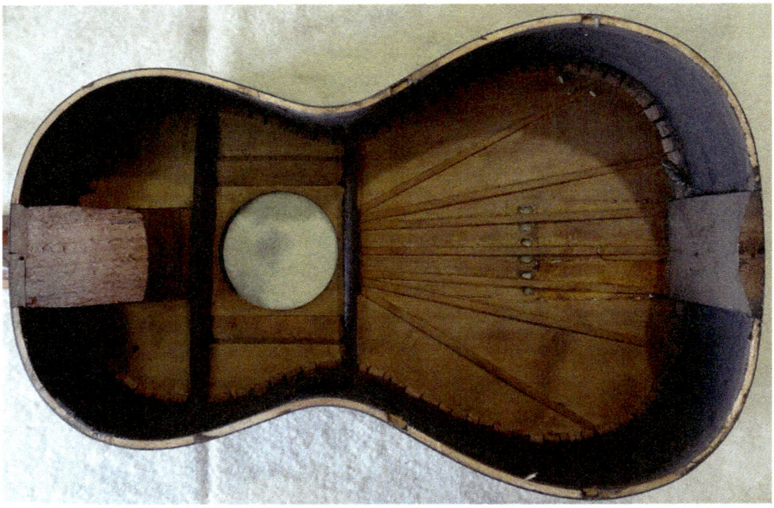

Fig. 2.6 Fan Bracing in an 1840 Panormo Guitar (image courtesy of Arthur Robb, art-robb.co.uk)

Figure 2.5 shows the ladder braced soundboard of a baroque guitar made by Emily Shaw, inspired by one from Antonio Stradivari (yes, that Antonio Stradivari, the violin guy). This picture also shows several light spruce reinforcement strips and the back of the soundhole decoration.

Guitars with six courses (pairs) of strings appeared in the middle 1700s. As the guitar developed, luthiers moved from ladder bracing to fan bracing. Pictures of soundboards of the earliest instruments braced this way are very rare; to show the idea, Fig. 2.6 shows the soundboard of a guitar made in 1840 by Louis Panormo.

Fig. 2.7 A guitar made by Francisco Sanguino in 1770 (image courtesy of the Musical Instrument Museum, Phoenix, Arizona, USA, mim.org)

This instrument underwent a careful restoration by Art Robb, who removed the back, exposing the inside of the soundboard. It's easy to see why this pattern is called fan bracing.

Figure 2.7 shows a guitar made by Francisco Sanguino in 1770. While it still bears a strong resemblance to the baroque guitar, it is a step toward modern instruments. It has six courses of paired strings and the soundhole is open—free of the parchment decoration that appeared on previous instruments.

Fig. 2.8 Bracing of Voboam Guitar, 1690–1699 (image courtesy of Richard Bruné, rebrune.com)

This particular instrument has a form of ladder bracing, though a similar instrument he made in 1759 may be the first guitar to use what we would now consider fan bracing. As mentioned earlier, there is no single path through the development of the guitar. So, it's risky to state that a specific instrument is the first example that contains some feature. In this case, an instrument by Alexander Voboam from around 1690–1699 has two small diagonal braces on the soundboard that are precursors to later fan braces (Fig. 2.8). Note also the diagonal brace in the lower bout in addition to the traditional ladder braces.

Another step on the way to modern guitar designs is the early romantic guitar. There is no clear definition of the early romantic period and this was not a name used at the time. But, based on the instruments more than the music associated with them, we might consider the early romantic period to run from the late 1700s to the mid-1800s.

Figure 2.9 shows one made by Jean-Nicolas Grobert around 1830. This guitar combines some features that seem more modern with some that clearly aren't. It has six single strings, fixed at the bridge with pins. The neck has fixed metal frets. Also, the body shape is closer to that of modern instruments and it is free of the elaborate, sometimes ostentatious decoration sometimes seen in earlier instruments.

Fig. 2.9 An early romantic guitar by Jean-Nicolas Grobert from about 1830 (Wikimedia Commons, commons.wikimedia.org, original instrument in the collection of the Philharmonic of Paris, collectionsdumusee.philharmoniedeparis.fr, original photo by Richard Lambert)

Still, there are some obvious elements that are clearly held over from previous instruments. The fretboard is level with the soundboard and stops at the ninth fret. The remaining frets are set directly into the soundboard, which extends onto the neck. Also, wood friction pegs are used to adjust string tension. There is still some distance between this instrument and modern ones.

Guitar designs evolved through the nineteenth century, introducing some now familiar features. Figure 2.10 shows the body of a guitar made in 1839 by Louis Panormo with a fretboard extending over the soundboard, as in modern instruments. The body is still quite narrow and the strings were originally gut (now nylon). James Westbrook notes an interesting detail of this instrument is that has two distinct layers of finish, visible under an electron microscope. Because of an incompatibility in the two layers, they have tended to delaminate over time, making the instrument look more worn than contemporary instruments by another builder.

Art can be a useful way to track the evolution of guitar design as artists often faithfully record the details of instrument played by a sitter. In 1871 or 1872, Edgar

Fig. 2.10 Body of an 1839 Guitar by Louis Panormo (image courtesy of Dr. James Westbrook)

Degas painted Lorenzo Pagans playing a romantic guitar. This image is a reminder that musical instruments often have long working lives and that new designs take time to spread. Degas painted this picture something like 15 years after the advent of the modern classical guitar in Spain (Fig. 2.11).

Before moving on, we need to know the names for the parts of a classical guitar (Fig. 2.12). These names are also used for the steel string guitars we'll see later.

2.3 The Modern Classical Guitar

For decades, the design of the classical guitar has been evolving and features that were once seen as experimental became more common. Even so, there is a guitar that is generally acknowledged as the first modern classical guitar, made in 1856 by Antonio de Torres Jurado. This guitar, called La Leona (The Lioness), incorporated a group of features that, collectively, mark it as the first modern classical guitar. These include what we now consider a full-sized body of the shape that is now nearly universal. It was larger than the Romantic guitars that preceded it, with a

Fig. 2.11 A painting by Edgar Degas showing a romantic guitar (Wikimedia Commons, commons.wikimedia.org, original painting is in the Musée d'Orsay)

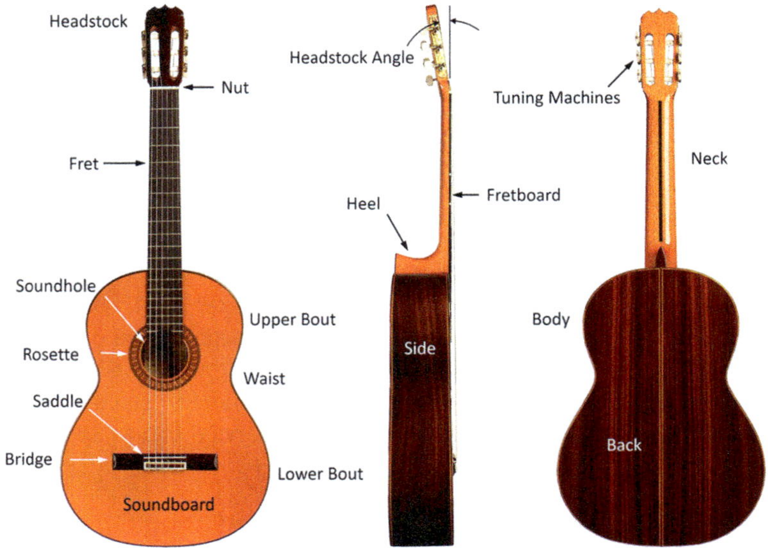

Fig. 2.12 Parts of a classical guitar (images of Ramirez 1A courtesy of Guitar Salon International, guitarsalon.com)

Fig. 2.13 A replica of the 1856 Torres La Leona (FE 04) by Fritz Ober (photograph by Felix Salazar, reproduced courtesy of Guitar Salon International, guitarsalon.com)

body length of 18.27 in. (464 mm). Modern full-sized classical guitars are usually a little larger still, with body length of 19.25 in. (489 mm) being typical.

The soundboard is fan-braced with seven bars (also sometimes called struts) arranged symmetrically around the centerline. The neck joins the body at the 12th fret, usually called a 12-fret neck. Jose Romanillos reports that this guitar originally had wood friction pegs for tuning, as was then common, and was later converted to accommodate geared tuners. Figure 2.13 shows a reproduction of the 1856 Torres made by Fritz Ober. The differences between this instrument and that by Grobert in Fig. 2.9 are clear. Torres' guitar appears to us a modern instrument while Gobert's, only 26 years older, doesn't.

Romanillos gave guitars made by Antonio de Torres unique identification numbers, beginning with either FE or SE. Torres made guitars during two distinct periods in his life. The first was when he worked in Sevilla, a time referred to as his first epoch (FE). The second was in Almería, a time referred to as his second epoch (SE). He appears to have stopped making guitars for several years in between these two periods when he operated a China shop with his wife. His guitars are, thus, identified by unique numbers and identified as being from either his first or second epoch. The designation FE 04 for his 1856 guitar indicates that it was the fourth guitar from his first epoch. Note that there is some question about Romanillos' attributions of instruments in the first epoch. It would be surprising, for example, that La Leona would be only Torres' fourth instrument.

The name classical guitar only became widely used after WWII. Before then and outside of Spain, it was generally called the Spanish guitar. A hint of this is in the earliest electric guitar introduced by Gibson. The ES-150 "Charlie Christian" is one of the first commercially successful electric guitars. The "ES" stands for "Electric Spanish." In keeping with modern practice, I will generally refer to Spanish guitars as classical guitars.

2.3.1 *Classical Guitar Dimensions*

Unlike other types of guitars, the dimensions and body shapes of classical guitars are nearly standardized. In this sense, they are somewhat akin to violins. While there is no one universal design, modern instruments closely follow the proportions of the 1856 Torres, whose dimensions appear in Table 2.1. Romanillos does not report the width of the nut, though modern copies typically have a 51 mm nut.

2.3.2 *Classical Guitar Materials*

Almost all classical guitars are made of wood and that is unlikely to change soon. Classical guitar design is strongly influenced by tradition and there are few alternative materials being used in classical guitars. The most traditional woods for classical guitars are:

Top	Spruce or cedar
Sides	Brazilian rosewood
Back	Brazilian rosewood
Neck	Spanish cedar or mahogany
Fretboard	Ebony
Bridge	Ebony

Note that, while several species of spruce are widely used, other conifers with similar mechanical properties have been used. This group includes larch and several species of fir.

Sometimes Indian rosewood is substituted for Brazilian rosewood and rosewood is sometimes substituted for the ebony. For example, Segovia's 1937 Hauser has an Indian rosewood bridge, though it has Brazilian rosewood sides and back. It is

Table 2.1 Dimensions of 1856 Torres classical guitar (FE 04, La Leona) as reported by José Romanillos

Dimension	Metric (mm)	English (in.)
Upper Bout	263	10.35
Waist	229	9.02
Lower Bout	343	13.50
Body Length	464	18.27
Scale Length	649	25.55
Depth (top)	91	3.58
Depth (waist)	94	3.70
Depth (bottom)	97	3.82
Soundhole Diameter	86	3.39

unknown whether his use of Indian rosewood for the bridge is intentional or a substitution. Modern luthiers now use a wider range of species for backs and sides.

Classical guitar tops are universally made from a small group of conifers (evergreens). Sometimes these are referred to as pine, though actual pine (genus Pinus) is seldom used. Some commonly used species, along with their formal names, are:

- Sitka spruce (Picea sitchensis).
- Engelmann spruce (Picea engelmannii).
- European spruce (Picea abies).
- Lutz spruce (Picea x lutzii).
- western red cedar (Thuja plicata).
- Redwood (Sequoia sempervirens).

Lutz spruce is a hybrid species that has only recently come into use. Western red cedar has only been used for classical guitars since the 1970s and redwood is used only rarely. Picea abies is also popularly called Norway spruce, European spruce, German spruce, and Carpathian spruce.

The Latin names are from the taxonomic system developed by the Swedish botanist, Carl Linnaeus, in the 1700s. He needed a way to organize the huge number of plants, so he developed a logical system of classification that we still use. If you wish to keep track, the taxonomic divisions are: domain, kingdom, phylum, class, order, family, genus, and species. Wood is typically identified by the genus and species. For example, Indian Rosewood is genus Dalbergia and species Latifolia. It can sound a little pedantic to talk about wood using the taxonomic names, but there is no other unambiguous way to do it.

A basic problem with discussing wood species is that they are often misidentified. Even experts sometimes identify wood samples incorrectly and it's prudent to assume that species listed by guitar makers and material suppliers are sometimes a little off.

There are several additional problems. The first is that a particular species with a single taxonomic name might have several common names. For example, Pau Ferro, a tropical hardwood used for sides, backs, and fretboards, is formally called Machaerium scleroxylon. However, it is commonly called Pau Ferro, Morado, Bolivian Rosewood, and Santos Rosewood. The last two names are used, even though Pau Ferro is not even in the same genus as rosewood. It doesn't help that some woods have many close cousins. For example, the genus Dalbergia that includes Indian and Brazilian rosewood includes dozens of species, only some of which are commonly called rosewood. Cocobolo and Brazilian tulipwood are both members of genus Dalbergia.

The various common names seem to have arisen naturally as different people needed names for wood they were using. However, sometimes the reasons are more commercial. For example, Gibson used balsa center blocks in semi-hollow body guitars. This is a very light wood, with some attractive mechanical properties (ask anyone who makes model airplanes), so it was probably a good design choice. However, they called it "Chromyte" in their marketing literature. This seems to be a made-up name, perhaps derived from the taxonomic name, Ochroma pyramidale.

The second problem is that there is no robust, practical way of identifying the species of a wood sample. Leaves and bark greatly assist in identifying trees, but boards of different species sometimes look enough alike to make positive identification a challenge. Even skilled people sometimes struggle to correctly identify wood samples. The problem is magnified by the range of species that have been used. Even the best luthiers have necessarily used whatever was available or came readily to hand. Actually, the problem is even worse than that because natural hybridization of species in plants is much more common than in most animals. For example, Lutz spruce is a naturally occurring hybrid of Sitka Spruce and White Spruce. Nobody bred Lutz spruce; it just appeared on its own.

Aging makes the problem worse and luthiers sometimes have stories of working on old instruments on which some wood is misidentified. Species are genetically unique and could, in principle, be reliably identified this this way. However, this is not yet practical. For now, lumber in the absence of bark or leaves is identified visually or by smell.

Of course, it's not clear how big a problem all this really is. A good guitar is just that, no matter what woods it uses. For many luthiers, an attractive guitar with a pleasing sound and a good feel is the only goal. Customers don't always see it that way, though, and correct labels can be important.

The combination of a spruce top, Brazilian rosewood sides and back, a Spanish cedar neck, and an ebony fretboard is one of the most traditional for classical guitars. However, this is changing as Brazilian rosewood and ebony have become scarce. Some woods have been over-harvested and are now protected by international agreements. At this writing the most important one for luthiers is CITES (the Convention on International Trade in Endangered Species of Wild Fauna and Flora). It is a wide-ranging agreement and guitars are only a minor part of its effects. Acoustic guitar manufacturing uses a vanishingly small portion of the wood harvested every year, so restricting species available to luthiers has almost no practical effect. However, it is not reasonable to expect a separate international agreement just for guitar wood. For now, CITES strongly affects the availability of wood for guitars.

Some simple estimates suggest how much wood acoustic guitar manufacturers might use and how tiny it is in comparison to other industries. A standard 20-foot freight container is 1 TEU or 20-foot Equivalent Unit and has an interior volume of 1280 ft^3 or 15,360 board feet. I know these are goofy English units, but that's how it's defined. If we generously assume that an acoustic guitar requires 4 board feet of wood, then a single container can hold enough wood for more than 3800 guitars. Thus, a single shipping container could hold enough guitar wood to supply a mid-size manufacturer for months. On a more personal scale, I could put more acoustic guitar wood in the back of a minivan than I'll use for the rest of my life. For further context, a modest container ship is about 5000 TEU and a really big one is 20,000 TEU.

While it is certainly true that acoustic guitar manufacturing has almost no effect on the world's forests, I would still much rather use sustainably harvested woods than worry that I'm treading too heavily on the Earth. Fortunately, formal experiments have shown that changing back and side materials has little discernable effect on sound quality—luthiers can use whatever materials they like for the back and sides. In response, a large variety of hardwoods are now offered by lutherie suppliers.

Figure 2.14 shows a ¾ size acoustic guitar made whose back and sides are made from Black Limba. Its scientific name is Terminalia superba, but it is commonly called Black Limba, White Limba, Korina, and Afara. The binding and end graft are from a walnut tree cut on the campus where I work, as part of the normal grounds maintenance. This is a successful instrument and the fact that Black Limba used to be considered almost a junk wood doesn't matter. Indeed, some potential buyers found the dramatic grain appealing.

Perhaps, the most famous experiment with alternative back and side materials is a guitar made by Torres in 1862 for which he made the back and sides from cardboard (it is sometimes described as papier maché, but this may be a quirk of translation). There is at least one recording of this guitar and it sounds good, with a pleasing, clear tone. A more recent experiment is a classical guitar made by Robbie O'Brien that uses heavy poster board for the back and sides (Fig. 2.15). It is a successful instrument and sounds fine.

The most common alternatives to wood are fiber composites such as fiberglass and carbon fiber, but they have made almost no inroad into classical guitar making. A few companies make steel string guitars from carbon fiber, but the classical world is still resistant. It's worth noting that there is at least one such successful manufacturer of instruments in the violin family. Yo Yo Ma famously plays a carbon fiber cello made by Luis and Clark, so it is risky to dismiss man-made materials for classical guitars. For now, though, classical guitars are made from wood.

A material that is easy to overlook is glue. Guitars were originally made with hot hide glue, though classical guitars and a few higher end steel string instruments

Fig. 2.14 A ¾ size acoustic guitar made with Black Limba and Walnut

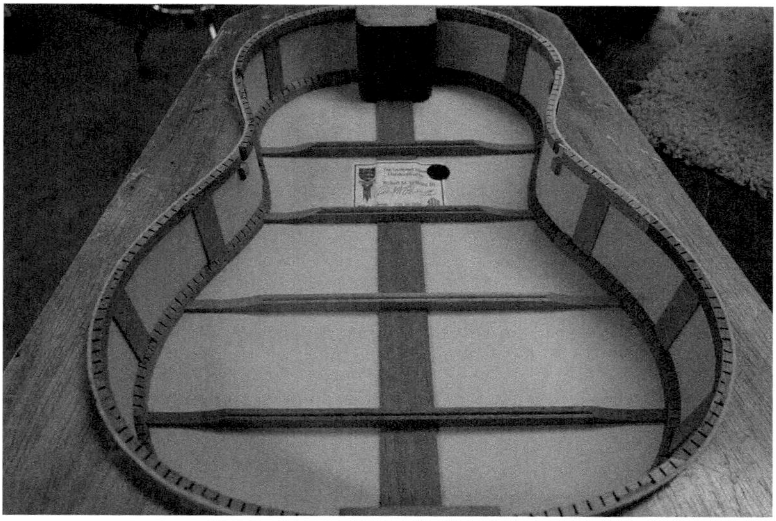

Fig. 2.15 A classical guitar by Robbie O'Brien with poster board back and sides (image courtesy of Robbie O'Brien, obrienguitars.com)

from manufacturers (e.g., Martin Authentic series) are the only ones still routinely made with it. Most acoustic guitars are assembled with yellow wood glue which is a synthetic polymer called aliphatic resin (AR). Hide glue is just what it sounds like. It's an organic colloid of protein made from animal hides. It has been used for millennia, so its properties are very well understood. It has also been shown to hold up well over very long times.

Hide glue has some very attractive mechanical properties. Its proponents state that it resists creep under load at room temperature, so joints move little over time, though it must be said that there is little supporting data in the technical literature. It's reversible and sticks to itself, so instruments made with it can be disassembled for repairs. It's plenty strong enough for guitars and dries hard so that it adds little damping to glued joints. Lutherie suppliers offer both granulated hide glue and liquid fish glue, a variation of natural protein glue sometimes used for repairs.

The disadvantages of hide glue are serious enough that it has largely been replaced with AR. Hide glue needs to be prepared before it can be used. It is supplied in granules that must be soaked in water and then heated to form a thick liquid. It must be kept warm since it solidifies as it cools. Open time is short and the reconstituted glue has a relatively short life unless refrigerated when not needed. When cured, hide glue is sensitive to heat and humidity, though it may not be more sensitive to heat than AR glues. The manufacturer of a popular AR glue states that overnight exposure to 150 °F (65.5 °C) lowers bond strength by 55%.

Some of the finest luthiers will not use anything but hot hide glue and only a little practice is needed to learn how to work with it—I only needed a couple interesting hours of gluing test blocks together to learn how to handle it. Instruments made with hide glue can be readily disassembled for repair, greatly extending their working life. All of the legendary classical guitars, some of which are still playable after more than 100 years, were made with hide glue.

2.3.3 Classical Guitar Structure

The structure of the traditional classical guitar is well defined. The instrument is made as a single component, with the sides set into the heel (Fig. 2.16). This distinguishes them from steel string guitars in which the neck and body are separate structural element that are joined late in the construction process.

Steel string guitars have stiffening rods in the neck called truss rods, usually with adjustable tension to change the curvature of the neck under string loads. Classical guitars seldom have adjustable truss rods and often have no truss rods at all.

Classical guitar backs have several lateral braces and the top usually has several lateral braces (sometimes called harmonic bars), placed on either side of the soundhole and accompanied by braces on the lower bout, usually arranged in a fan pattern. Classical guitars usually have reinforcing plates on the inside of the top, called bridge plates, while their closely related cousins, flamenco guitars, do not.

A seven-bar fan bracing pattern became widely used by the turn of the twentieth century and is still the most popular bracing pattern in classical guitars. As example, Fig. 2.17 shows a plan of an 1888 guitar by Torres (SE 114). After restoring this instrument, Jeff Elliott drew this detailed plan which is available from the Guild of American Luthiers (GAL Plan #52–1888 Antonio de Torres Classic SE114). The seven fan braces are placed symmetrically with respect to the center line, with three transverse (harmonic) bars at the waist and upper bout. It also has two additional bars, arranged in a shallow V and the ends of the fan braces. Note that, as was common in nineteenth century Spain, this instrument has no bridge plate.

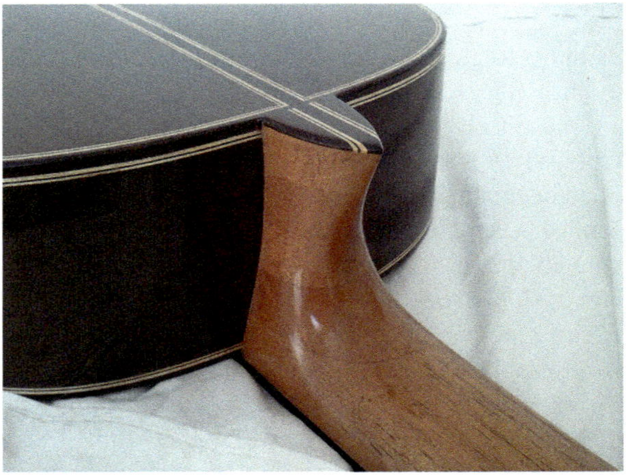

Fig. 2.16 The heel of a classical guitar

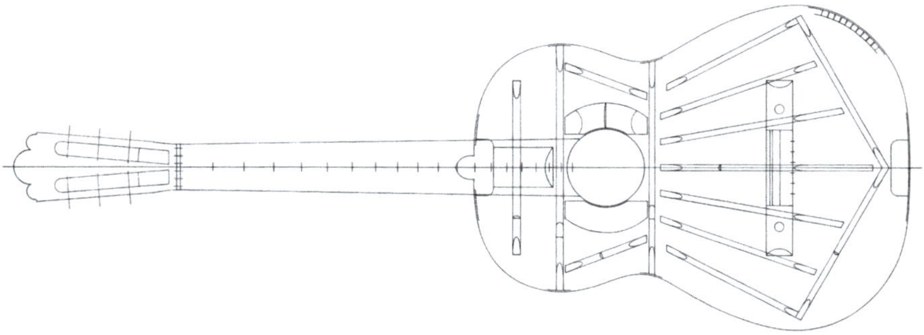

Fig. 2.17 A plan of the 1888 Torres SE114 (courtesy of the Guild of American Luthiers, luth.org)

Fig. 2.18 A guitar with traditional fan bracing by Jeff Elliott (image courtesy of Jeff Elliott, elliottguitars.com)

A very clean example of the Torres/Hauser bracing pattern is shown in a guitar by Jeff Elliott (Fig. 2.18). Note that this instrument has a bridge plate.

While many structural elements are hidden from the player, one that is not is the relationship between the neck and body. Traditionally, the top of the neck (bottom

Fig. 2.19 Traditional neck and body geometry for a classical guitar (image courtesy of Guitar Salon International, guitarsalon.com)

Fig. 2.20 A flamenco guitar under construction, showing the joint between the sides and neck

of the fretboard) has been even with the body at the joint and with the headstock slightly elevated as shown in Fig. 2.19. This is a fine classical guitar made by Herman Hauser II in 1957 and owned by Julian Bream. The superimposed dotted line shows that the nut is slightly raised above the plane of the top. There is no standard dimension, but it is typical for the nut to be raised 2–4 mm (0.079–0.157 in.). This allows the bridge to be slightly lower, reducing its torque on the top, while not bringing the strings too close to the fretboard or top.

The slightly raised neck angle is easy to implement with the traditional construction method in which the sides are set into slots cut directly into the heel. Figure 2.20 shows a flamenco guitar made by the author in a class taught by Paco Chorobo. The slots in the unshaped heel are clearly visible.

Fig. 2.21 A 1994 Humphrey Millennium model with an elevated fretboard (image courtesy of Guitar Salon International, guitarsalon.com)

While most classical guitars follow traditional patterns, it's quite wrong to think that modern builders are in stasis. Some classical guitar makers are in a state of continuous experimentation in an effort to refine their designs. Some of their refinements are visible to the player and some are not.

A different take on the alignment between the neck and body is an elevated fretboard, a refinement intended to improve access to the higher frets. This design has been adopted by some top luthiers. Figure 2.21 shows a Millennium model classical guitar made in 1994 by Thomas Humphrey. The body is clearly tapered with the fretboard raised well above the top.

The history of the classical guitar is rich with experimentation and it can be difficult to find a "new" design element that is truly new. Richard Bruné observes

> Most innovations are more accurately refinements or revisiting of ideas that were previously tried. For instance, the elevated fingerboard is a device that harks back to the early 19th century Italian model of Luigi Legnani which became popular in Vienna and German speaking areas when he gave concerts on his new design.

Reaching the higher frets can be difficult on a traditional classical guitar. Designers of steel string guitars addressed the problem of neck access long ago with cutaway bodies being in production at least since the early 1900s. Figure 2.22 shows a cutaway on a Taylor model 214ce-N nylon string guitar. Experiments with cutaways certainly predate production models by decades. However, classical guitars with cutaways remain rare. While the Taylor does have nylon strings, it is not built on the pattern of a classical guitar and is not intended for traditional classical players.

A variation is a partial cutaway that improves neck access with less change in body volume than a full cutaway. Figures 2.23 and 2.24 show a classical guitar made in 2013 by Richard Bruné, a top American luthier and historian of the classical guitar. It has a partial cutaway that includes an acoustic port in addition to an elevated fretboard. Bruné calls this partial cutaway a "biteaway."

Richard Bruné on Design

After the touch of the player, the design is what makes a guitar sound the way it does. Wood in and of itself has no inherent "sound." Only until you shape it into some very specific form and vibrate the attached strings does it begin to produce sound. So, one could make a Martin D-28 and a 1937 Hauser copy

Fig. 2.22 A Taylor model 214ce-N with a cutaway upper bout (image courtesy of Taylor Guitars, taylorguitars.com)

Fig. 2.23 A classical guitar made in 2013 by Richard Bruné with ported partial cutaway and elevated fretboard (image courtesy of Richard Bruné, rebrune.com)

Fig. 2.24 A ported partial cutaway in a guitar made by Richard Bruné in 2013 (image courtesy of Richard Bruné, rebrune.com)

using matching soundboards but they would have totally different characteristics as musical instruments because their design is radically different. This principle applies to more subtle differences between similar designs such as a Hauser vs a Velazquez guitar which to the average person appear to be nearly identical. Any guitar is a summation of all its design details. Getting these details right is more important than the wood selection. The best wood in the world will never overcome a bad design.

Other classical builders have incorporated acoustic ports in their instruments and in a range of different geometries. Figure 2.25 shows the upper bout of a guitar made by Petr Matousek in 2008. This fine instrument incorporates a pattern of differently sized ports in the upper bout. Note the elevated fretboard. Also, just visible at the edges of the ports are the laminations that form the sides.

Laminated sides, as distinguished from industrially manufactured plywood, are an increasingly common feature in classical guitars, though they are hardly new. They typically have two or three layers, with the outer layer often being a traditional hard wood, such as rosewood. The inner layer might be something softer and less visually interesting. For example, John Bogdanovich often uses yellow cedar or Sitka spruce in 0.6 mm (0.022 in.) veneers in two layers for the inner plies. Laminated sides are more resistant to cracking than solid wood sides.

One of the most important areas of experimentation among classical guitar makers is in bracing. It is structurally possible to make a classical guitar with no bracing

Fig. 2.25 Sound ports in a guitar made by Petr Matousek in 2008

other than the transverse braces on either side of the soundhole. Indeed, early guitars were often made without bracing on the lower bout. One of the few modern builders who have made a guitar this way is Richard Bruné, who made a cypress flamenco guitar with a 2 mm (0.079 in.) cedar top that had no fans, only transverse braces above and below the soundhole. At this writing, the instrument is 23 years old and still functioning well. While Bruné describes its sound as bland, Pepe Romero played it at the 1997 Guitar Foundation of America Festival. Bracing is not structurally necessary for a classical guitar but, rather, is important in refining the sound of the instrument.

Most classical guitars still use some variation on the fan bracing used by Torres in his 1856 instrument and by Hauser in his 1937 instrument. For example, Fig. 2.26 shows the interior of an instrument made by Hermann Hauser in 1930. Note the numerous cleats used to repair cracks in the top. Bruné made a careful restoration that required removing the back and exposing the bracing.

Fig. 2.26 A fan braced guitar made by Hermann Hauser in 1930 (image courtesy of Richard Bruné, rebrune.com)

It is important to note that the bridge is an important structural element on the soundboard. It is very stiff in comparison to the top and the braces, and certainly affects the sound of the resulting instrument. For example, Bruné notes that the wings of the bridge on the 1937 Hauser are heavier than one might expect and suggests that this may be important in forming the tone of the instrument.

Variations in bracing range from subtle to extreme. As an example of a successful refinement, Jeff Elliott, a top builder who has Julian Bream among his clients, incorporated openings in the three lateral or transverse bars (also called harmonic bars) at the waist and upper bout, among other features (Fig. 2.27). The idea of open harmonic bars extends back to at least Torres, as evidenced by his famous 1856 guitar (FE04 La Leona), which has one open lower transverse bar. The soundboard is supported by a metal tornavoz (an internal soundhole extension) that rests on small wood supports fixed to the back.

Open transverse bars are sometimes found in other mid-nineteenth century guitars that also appear to have been originally fitted with a tornavoz. As an example of this device, Fig. 2.28 shows a tornavoz in a 1939 Llobet guitar by Hermann Hauser Sr., which was undergoing restoration by Richard Bruné when this picture was taken.

Fig. 2.27 A classical guitar with open harmonic bars (image courtesy of Jeff Elliott, elliottguitars.com)

Fig. 2.28 A 1939 guitar by Hermann Hauser Sr. with a tornavoz (image courtesy of Richard Bruné, rebrune.com)

Fig. 2.29 An open harmonic bar with a 0.5 mm cross grain reinforcement strip (photo by Mike Doolin, doolinguitars.com, courtesy of Jeff Elliott, elliottguitars.com)

While the use of a tornavoz has since been largely discarded, the open harmonic bar has remained as a design feature. Elliott has worked to refine the idea, inspired by Julian Bream's 1973 José Romanillos guitar. A detailed plan of his design is available from the Guild of American Luthiers (Plan #74—Jeffrey R. Elliott Open Harmonic Bar Classic). Elliott feels that

> the purpose of the open harmonic bars is to allow the surfaces of the waist and upper bout to work in consort with the lower bout, so that the entire top responds more fully, and all together. However, there is both a sonic and structural risk inherent in the design: The trick is to control the responses so that they don't cloud the fundamental, while also providing enough structure to avoid distortions and cracking. When executed well, the design provides for a more immediate response to touch, an overall increase in volume and projection, a broader tonal and dynamic range, more upper-partial and harmonic responses with greater sustain, and a more mature sound as a new instrument.

Open transverse bars eliminate some of the cross-grain reinforcement, possibly allowing cracks to form in the top. Elliott has seen this failure in a few older guitars with open transverse bars. To address this possible failure, he adds a 0.5 mm (0.020 in.) reinforcement to the top, below the open portion of the braces. This subtle, but important structural element is visible under the open bars in Fig. 2.29.

My Principal Considerations in Classical Guitar Design

By Jeffrey R. Elliott

When building for a client, I first establish a common language to describe the desired sound quality and character. If possible, I listen to and watch the client play, and we listen to some of his/her favorite guitar music together. The terms we each use to describe the sounds we're hearing don't have to be the same, so long as *we both hear the same thing* when we use those words. Once we prioritize our acoustic goals and define our expectations, I determine the primary purpose for the instrument's use (genre, style of playing, playing technique, recording, amplification, and so on), and if touring and road-worthiness is a consideration, I pay particular attention to structural stability. Finally, we discuss and decide the physical requirements, including string scale, string tension, nut and saddle string spacings, neck shape, fret details, playing action, ergonomics, electronics, and so on.

I believe that the primary factors that account for what I hear in my instruments are a combination of the tonal properties of the woods, the top's bracing design, and my own "touch" as the maker. I'm an intuitive builder, and when I'm on the fence with a decision or am in doubt, I try not to second-guess myself.

For me it's both practical and enjoyable to make detailed drawings before beginning the construction. Typically, I draw a full-scale set of plans that include both the top view, and perhaps more importantly, the side view. The top view shows the instrument's centerline, bridge placement, and the location of the 12th fret/neck/body joint. The drawing so far is home base, and everything else revolves around it. The side view shows the plane/angle of the neck/body joint, the fingerboard thickness/taper, fret height, the playing action (showing the strings and the daylight between them and the frets), the arching in the top, and the string height off the top at the bridge. The geometry is very sensitive, and if one of these variables changes even a little, it can disproportionately affect the end result, so I must reconsider all of them before proceeding.

Finally, aesthetic considerations are important to both reinforce the player's bonding with the instrument and for the maker's satisfaction. And while individual touches can impact this greatly, function and comfort shouldn't be compromised for individual style.

A final observation on classical guitars on the Torres-Hauser pattern comes from Jeff Elliott, who notes that guitars using it have proven to be surprisingly durable. There are many instances of these instruments holding up under heavy use and even serious mistreatment. It seems obvious in retrospect that an instrument design can only become iconic if, along with its other merits, it is durable enough to become widely familiar.

Some builders have experimented with placing the braces in a lattice pattern to distribute the stiffness more evenly across the lower bout, and at least one manufacturer produces a lattice braced guitar. Figure 2.30 shows a Cordoba C12 with interior lighting to highlight the lattice bracing.

A more radical experiment in classical guitar bracing is from Australian luthier Greg Smallman, who developed a lattice bracing pattern that works in conjunction

Fig. 2.30 Lattice bracing in a Cordoba C12 (image courtesy of Cordoba Guitars, cordobaguitars.com, photo by Felix Salazar)

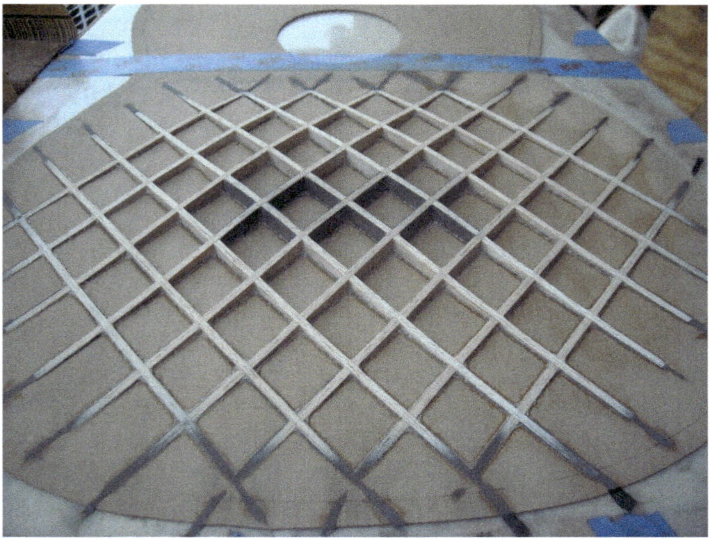

Fig. 2.31 A lattice braced top by David Schramm (image courtesy of Guitar Salon International, guitarsalon.com, schrammguitars.com)

with a heavy supporting structure. He replaced the small number of symmetric fan braces with a more distributed pattern of interlocking braces. Variations on this lattice pattern are used by some luthiers around the world. Figure 2.31 shows a partially completed lattice-braced soundboard by David Schramm, who is careful to

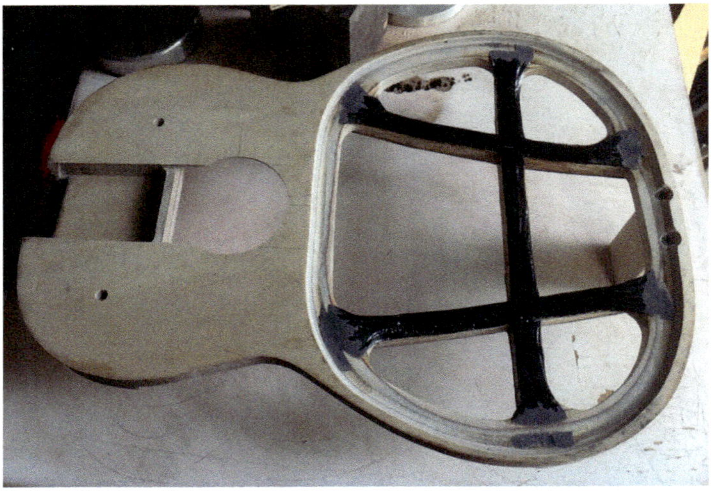

Fig. 2.32 Frame of a lattice braced guitar made by David Schramm (image courtesy of David Schramm, schrammguitars.com)

acknowledge the strong influence of Smallman in his work. The balsa lattice is eventually reinforced with unidirectional graphite strips.

A feature of guitars made on the Smallman pattern is that the body is heavily reinforced with an interior plywood frame to bear loads that would otherwise be borne by the top (Fig. 2.32). Note graphite reinforced arches that stiffen the structure without contacting the soundboard. It is clear from this picture that only the lower bout of the soundboard is free to vibrate.

The back and sides of Schramm's lattice braced guitars are also heavier than is traditional. Sides are laminated with several layers for a total thickness of about 3 mm (about 1/8 in.). The back is similarly laminated using veneers in alternating directions for a final thickness of 6–6.5 mm (about ¼ in.). Figure 2.33 shows the soundboard and frame glued to the laminated rim.

One last variation on top bracing worth exploring is double tops. Engineers have long known that the outer surfaces of plates and beams contribute disproportionately to their stiffness. Aircraft, where strength and weight are particularly important, sometimes use sandwich structures in which load bearing skins are bonded to a light core. One famous example is the de Havilland Mosquito (Fig. 2.34) that used plywood skins bonded to balsa cores. This strong, light structure contributed to its outstanding performance.

While not as flashy, the modern double top guitar uses the same structural concept. Typically, a thin layer of Nomex honeycomb is bonded between two layers of

Fig. 2.33 The interior frame of a lattice braced guitar by David Schramm (image courtesy of Guitar Salon International, guitarsalon.com, schrammguitars.com)

Fig. 2.34 A de Havilland Mosquito in USAAF Colors (USAF Museum, Wikimedia Commons, image is in the public domain)

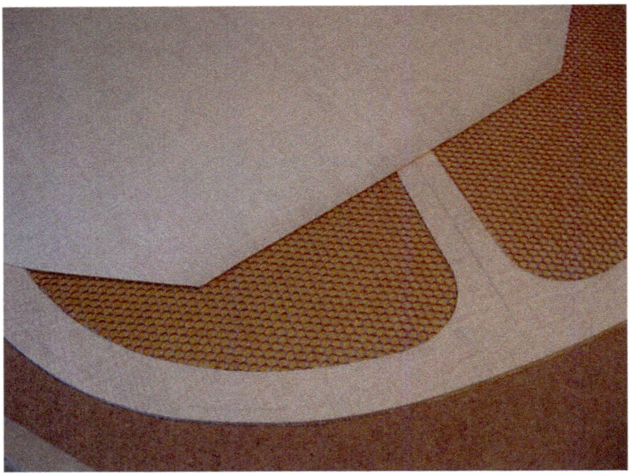

Fig. 2.35 Wood and Nomex layers for a double top (image Courtesy of Marcus Dominelli, dominelliguitars.com)

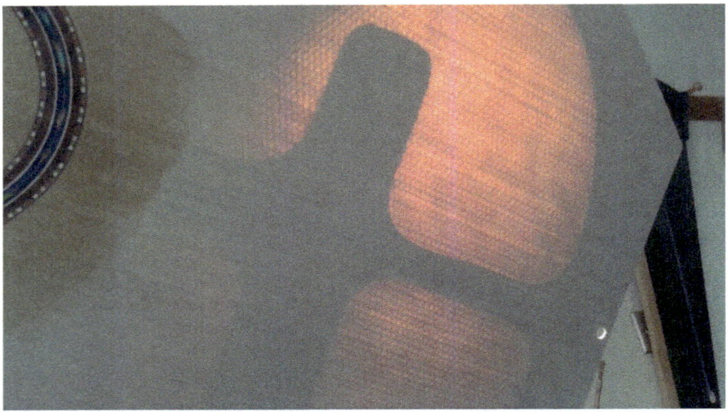

Fig. 2.36 A double top soundboard illuminated from within (image Courtesy of Marcus Dominelli, dominelliguitars.com)

spruce. The result is a light, stiff top that is attractive to some builders. Figure 2.35 shows the two spruce layers with the inset honeycomb core. One limitation of sandwich structures is that they aren't able to bear concentrated loads. Thus, guitar makers usually use solid wood at the edges and under the bridge. Here, one wood layer is thicker and has a shallow pocket to accept the Nomex core. The layers are very thin and it's typical for the Nomex core to be in the neighborhood of 1.5–2 mm (0.060–0.80 in.).

The thinness of the wood plates is shown in Fig. 2.36. A light bulb placed behind a finished top clearly shows the Nomex layer as well as the places that were left solid.

Not all innovations are flashy or even readily apparent. Richard Bruné posits that the patron saint of luthiers could be Willis Carrier, who developed the first practical air conditioning. The ability to control temperature and humidity has extended the working life of wood guitars to something akin to the working life of a guitarist. He says that before WWII, a large majority of guitar repairs were related to humidity variations during construction, but.

> With proper care, a guitar assembled under a single humidity equilibrium will never need repair due to humidity issues

Bruné's other choice for patron saint is Julian Hill, whose team at DuPont developed nylon in the 1930s. The gut strings that preceded them were expensive and not very durable. Evidence suggests there was a time when the cost of a year's supply of gut strings could be on the same order as the cost of the instrument. Cheap, durable nylon completely changed the economics of playing the classical guitar, making it much more accessible.

2.3.4 Classical Guitar Finish

Once the guitar is complete, it needs a protective finish. Both formal testing and common experience have shown that a heavy finish can affect the sound quality of even a very good guitar. Thus, the finish needs to be thin, light and have low damping, all while protecting the wood underneath.

Traditionally, the choices were limited and most guitars, no matter the type, were finished with shellac, varnish, oil, or lacquer. Shellac and lacquer dry by evaporation, so there is no chemical reaction. Varnish and oil undergo a chemical reaction while drying so that the cured finish is chemically different than what was originally applied.

Shellac is at least 3000 years old. It is made from the resin of the female lac bug, processed into flakes and dissolved in alcohol so that it can be brushed, sprayed on, or applied with a pad (French polishing). When the alcohol evaporates, the finish that remains on the guitar is still shellac, chemically unchanged from the original flakes, and can be softened again by applying alcohol. Because of this, the alcohol in successive layers partially dissolves the previous ones so that new layers "melt" together with the old ones. This is also sometimes called burn-in.

French polishing is a traditional finishing method that is still sometimes used on classical guitars. It applies shellac in a very thin layer using a cotton pad with a small amount of shellac, sometimes with other additives. While time-consuming, it produces a thin finish with a nice shine. Figure 2.37 shows Cyndy Burton French polishing a classical guitar with a small pad.

Fig. 2.37 Applying shellac by French polishing (image courtesy of Cyndy Burton)

Cyndy Burton on French Polishing

French polishing is sometimes described as a method for applying shellac. However, I tell my students that it's both a process and a finish. Sometimes it's called spirit varnish because many ingredients in addition to shellac (like copal and sandarac) are quite common, if you're making your own. I use 190 proof grain alcohol (ethanol) for the spirit part, because it is nontoxic. Although many polishers use a nondrying oil as a lubricant when applying, I don't use oil at all, and that makes the process more predictable. It's applied with a cloth-covered pad called a "muñeca." My favorite shellac is in flake form when purchased, and the raw materials it's made from are imported from India or the Far East for further processing in Germany.

There is a mystique about it, and like most things, you can make it very complicated if you want. Jeff [Elliott] and I feel very strongly about the physical beauty of a good French polish finish as well as its positive influence on the sound of the guitar, especially the heart of it all, the top. It's also the easiest finish to touch up and repair. It's more difficult to describe the process than to do it, and it's rewarding on many levels. I like to paraphrase Eugene Clark who said, "If you French polish, you'll have no need to go fly fishing." It takes focus, time, and patience, but it can be a peaceful, pleasurable experience.

The question often arises regarding French polish durability compared with other finishes such as nitrocellulose or water-based lacquer, polyurethane, polyester, and so on, especially for instruments other than classical guitars. For us, the positive qualities vastly outweigh the negative qualities of most other finishes, particularly those that are toxic for both the maker and the environment.

Classical guitarists often prefer instruments with French polished soundboards, presumably because the thin finish adds little mass or damping. However, the result is less durable than a modern polymer like polyester that must be treated with more care. Since shellac dries by evaporation, French polish finishes are not difficult to touch up or repair. Luthiers sometimes French polish soundboards while applying a more durable finish to the rest of the instrument.

The other traditional finish for classical guitars is varnish. Varnish is made from a drying oil, resins, and a thinner or solvent. The specific components vary quite a bit and violin makers, especially, sometimes have proprietary formulations. The drying oil was a plant or nut oil, such as linseed oil, tung oil, or walnut oil. Drying oils cure by a chemical reaction with the oxygen in the air, so the cured finish on the instrument is chemically different from what was originally applied. Drying oils like linseed (also called flaxseed) polymerize into a solid form and can be used to finish wood by themselves. Commercially produced oils often include drying agents to speed curing. Colloquially called Japan drier, these are generally salts of cobalt, manganese, or iron.

Production instruments generally use more modern finishes such as catalyzed polymers. Under controlled factory conditions, they can be applied thinly and uniformly enough to provide a durable finish without an unacceptable effect on sound quality. A few luthiers have experimented with removing some of the thick finish on the soundboards of some factory guitars in an attempt to improve sound quality.

2.4 Early Steel String Guitars

Steel guitar strings became commercially available in the US around 1900. While there were many guitar makers at the time, Martin is now considered the most influential. Martin Guitars didn't list steel stringed guitars in their catalog until around 1920, though small numbers of them appear in company records as early as 1900.

The development of the steel string guitar was gradual and evolutionary. Early American guitars unsurprisingly resembled contemporary European ones made for gut strings. Gut strings are just what they sound like, being made from natural fibers

Fig. 2.38 An 1836 Martin on the Viennese pattern (image by the author, reproduced courtesy of George Gruhn, guitars.com)

found in the walls of animal intestines. In the mid-1800s Torres, the great Spanish luthier, significantly improved the design of what we now call classical guitars, but his designs spread only slowly through the rest of Europe.

The American steel string guitar developed from the northern European gut stringed guitar. The designs of steel string guitars were driven by the higher tension of steel strings and the heavier structures needed. Christian Frederick Martin (C.F. Martin) was at the center of the development of the steel string acoustic guitar in the United States. He was born in 1796 and may have apprenticed with the Viennese luthier, Johann Stauffer. Martin company histories state that C.F. Martin apprenticed with Johann Stauffer, but there is no written record of it—the only extant document bearing both their signatures regards a patent claim and Martin signed only as a witness. It would be most unusual at that time and place for an apprenticeship to be undocumented.

Stauffer made guitars in a style that is still sometimes called a "Viennese guitar." Martin's early instruments drew heavily on Stauffer's designs. Figure 2.38 shows an 1836 Martin that shows the Stauffer influence.

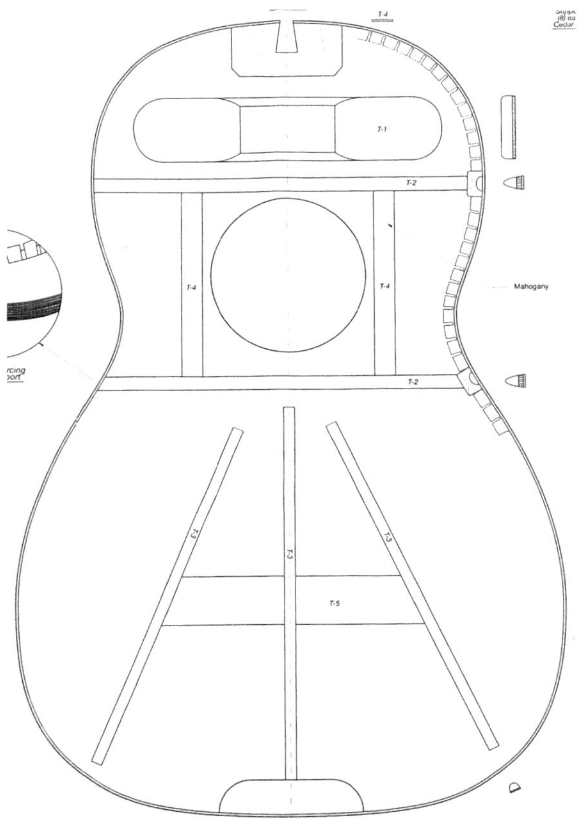

Fig. 2.39 Bracing pattern of a Martin Spanish Style Guitar c. 1842-43 (image courtesy of Stewart MacDonald, stewmac.com, plan drawn by Don MacRostie)

After a guild dispute over who should have the right to make guitars—something unknown in the US at the time—Martin moved to New York and then to Nazareth PA in 1839. His early instruments followed European designs, though with his own refinements. Martin's designs evolved in at least two different ways. One was the overall shapes and sizes and the other was the structure.

Structurally, three of the most important features on an acoustic guitar are the bracing pattern, the neck joint, and the headstock joint. Martin designs have become a touchstone for all three of these elements, even though it was only one of many companies making guitars in the United States. Martin's X brace pattern has become a de facto standard, so it's a good place to start.

Martin's early bracing patterns usually followed traditional designs. As an example, Fig. 2.39 shows the bracing pattern of a Martin Guitar circa 1842–43. Don MacRostie drew this plan after careful study of the original instrument. The bracing is a simple three-bar fan pattern, related to the seven-bar fan pattern used by Torres and Hauser.

Fig. 2.40 The de Goni 1843 Martin Size 1 with X bracing (courtesy of C.F. Martin & Co. Archives, martinguitars.com)

Martin guitars eventually became known for a bracing design called X bracing, that went on to become popular to the point of being ubiquitous. By the mid-1800s, Martin guitars had an early version of X bracing. Figure 2.40 shows a Martin Size 1 guitar from 1843 that uses an early version of it. The Size 1 is small by modern standards. It has a body length of 18 7/8 in. (479.4 mm), a lower bout width of 12 3/4 in. (323.9 mm), and a scale length of 24.9 in. (632.5 mm).

The guitar was made for the guitarist, Madame Delores N. de Goni, after she played an instrument by C.F. Martin that she found to be superior to her own Spanish style guitar. Figure 2.41 shows a daguerreotype of her with her Martin Guitar. Note that daguerreotypes laterally reverse the images and de Goni did not play left-handed.

Fig. 2.41 A daguerreotype image of Madame de Goni with Her Martin Guitar (image courtesy of C.F. Martin & Co. Archives, martinguitars.com)

Figure 2.42 shows a modern reproduction by the Martin Custom Shop of the top of de Goni guitar. The transverse brace above the soundhole and the X brace are immediately familiar. The lighter braces above the main X braces aren't part of later Martin designs and it's missing the smaller additional braces now universal in X-braced guitars.

Those interested in the history of the X-brace pattern note that a nearly identical pattern was used by Schmidt and Maul in the 1840s. The two contemporary companies, started by German immigrants and operating in New York, used similar bracing patterns. It's unclear which one developed this bracing pattern or whether one of them copied it from the other. Certainly, Martin popularized it.

The next important structural feature is the joint between the neck and the body, commonly referred to as the neck joint. Martin guitars, even the early one shown in Fig. 2.40, used a tapered dovetail joint. It was easy enough to make and strong enough to hold up, even under heavy use. Even now, many acoustic guitars use this type of joint. Figure 2.43 shows a neck and matching heel block with a dovetail joint, offered to luthiers by Stewart MacDonald. The joint is similar to the Martin design, though the slot for the truss rod and corresponding access hole in the heel block are more recent additions.

Fig. 2.42 A Martin Custom Shop Reproduction of the 1843 de Goni Guitar (image courtesy of C.F. Martin & Co. Archives, martinguitars.com)

Fig. 2.43 A dovetail neck joint (image courtesy of Stewart MacDonald, stewmac.com)

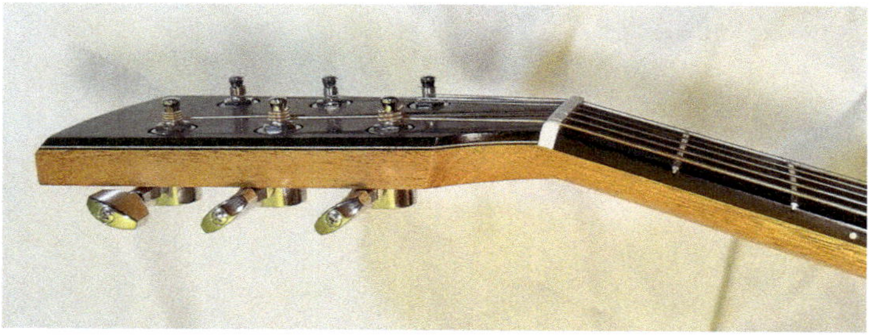

Fig. 2.44 The angled headstock on a steel string acoustic guitar

Fig. 2.45 A slotted headstock on a classical guitar by Kathrin Hauser, 2014 (image courtesy of Guitar Salon International, guitarsalon.com)

The third important structural feature is the joint between the neck and headstock. On almost all acoustic guitars, the headstock is at an angle to the neck. There is no standard angle, but 10–15° is typical. The angled headstock forces the strings to bend down as they go over the nut, which creates a downward force against the nut. The angle made by the strings is called a break angle and mechanically links the strings to the neck. Figure 2.44 shows the angled headstock on a guitar made by the author.

Headstock angles of classical guitars with slotted headstocks and geared tuners are lower (more parallel to the neck) since the string joins the tuner below the top surface of the headstock. This gives a larger angle of the string over the nut (called break angle) for a given headstock angle (Fig. 2.45).

Fig. 2.46 Disassembled headstock joint on an 1887 Martin Model 2½ -40 (image courtesy of Guild of American Luthiers, luth.org, original image by Frank Ford)

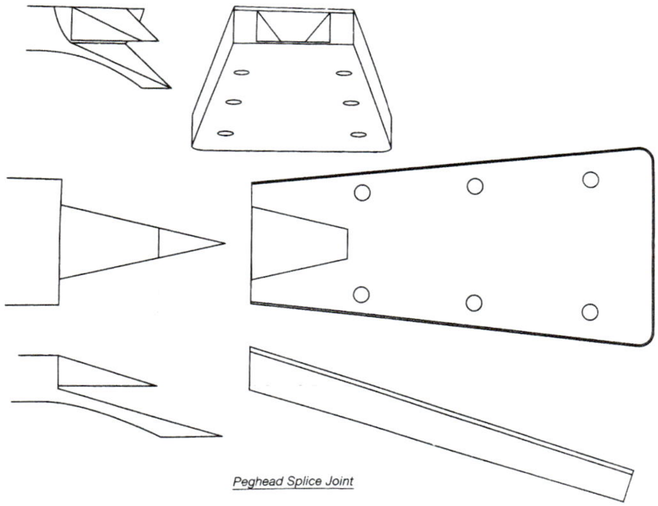

Fig. 2.47 Headstock joint detail of a Martin Spanish Style Guitar c. 1842-43 (image courtesy of Stewart MacDonald, stewmac.com, plan drawn by Don MacRostie)

Acoustic guitar necks and headstocks are usually assembled from straight pieces of wood, so there must be a joint between the neck and the angled headstock. Early Martin guitars used a very recognizable one with a diamond-shaped protrusion on the back called a volute. Figure 2.46 shows the disassembled neck joint of a Martin model 2½ from 1887. Glue is strongest in shear and this design has plenty of shear area.

This joint is also used on the c. 1842-43 Martin guitar plan by Don MacRostie. A detail from the plan is shown in Fig. 2.47. Some modern Martin guitars still have

Fig. 2.48 A scarf joint connecting the headstock to the neck

this volute between the neck and headstock, though this is often a decorative feature and not part of the neck joint.

This joint is strong and has some aesthetic attraction, but is difficult to make. A much more popular solution is a simple scarf joint, as shown in Fig. 2.48. It is easy to make and provides a large shear area for the glue, which is squeezing out from this freshly clamped joint.

There are two common ways to scarf join the headstock. One is to make the neck from one piece, with the scarf joint on the headstock, as shown in Fig. 2.48. The other is to make the headstock from one piece so that the scarf joint is on the neck. Figure 2.49 shows the two different ways to scarf join the headstock to the neck. The second design, with a two-piece neck, is also called a Spanish luthier scarf joint. Type 1 doesn't appear to have any common name.

As long as they are well made, both of these joints are strong enough to bear string loads along with a little heavy handling. I've been unable to find a clear history of where the two types of scarf joints originated. Neither does there seem to be common agreement about the merits of one over the other.

A consideration for the Spanish luthier scarf joint is the angle at which the glue seam intersects the neck. It is difficult to fair in a glue seam when it makes a small angle with the surface (Fig. 2.50). Without careful design, this angle can be quite small, making the glue seam more visible.

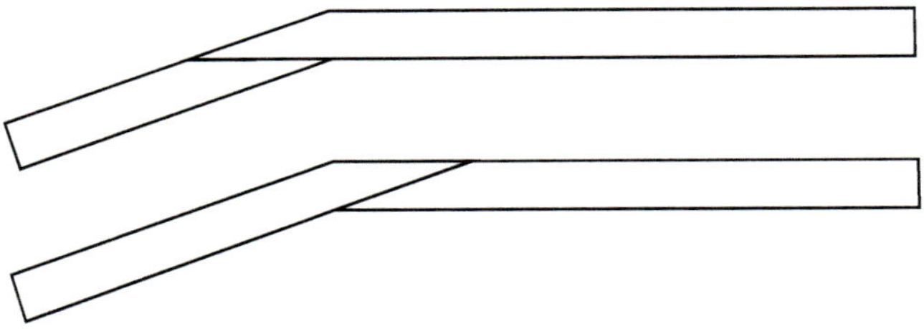

Fig. 2.49 Two different ways to scarf join the headstock to the neck

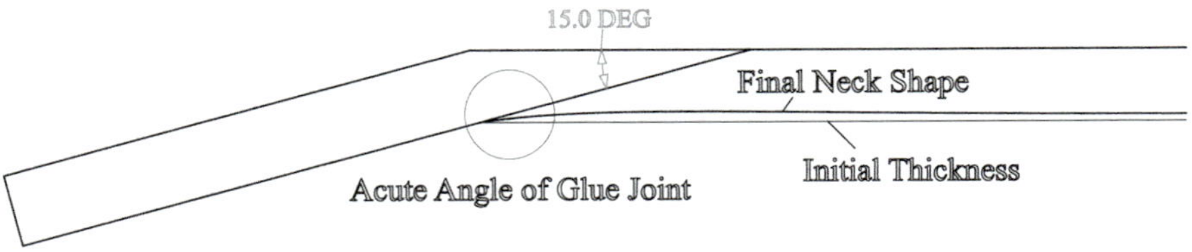

Fig. 2.50 Acute angle in glue layer of headstock scarf joint

Fig. 2.51 A volute in the headstock joint of a small acoustic guitar

One way to address this problem is to carve a volute into the joint, as shown in Fig. 2.51. This one is rounded, so that it is comfortable for the player's left hand. This instrument also has a thin plate on back of the headstock.

For all this, builders' choices of scarf joints vary. Type 1 scarf joints appear to be more popular, though the Spanish luthier scarf joint (Type 2) is also commonly used. Figure 2.52 shows the neck of a Fender CD-60S guitar with such a headstock joint, which also appears on their CC-60S. It's unusual for a scarf joint to be this far up the neck and suggests the blank for the headstock was much thicker than is common.

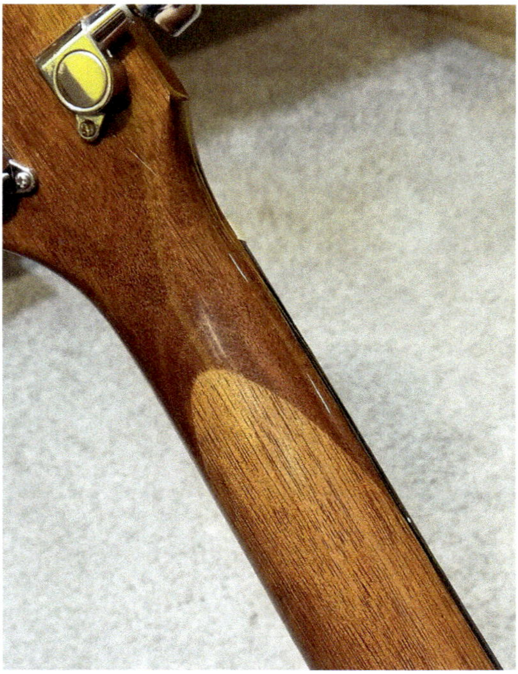

Fig. 2.52 Scarf joint on the neck of a Fender CD-60S dreadnought guitar

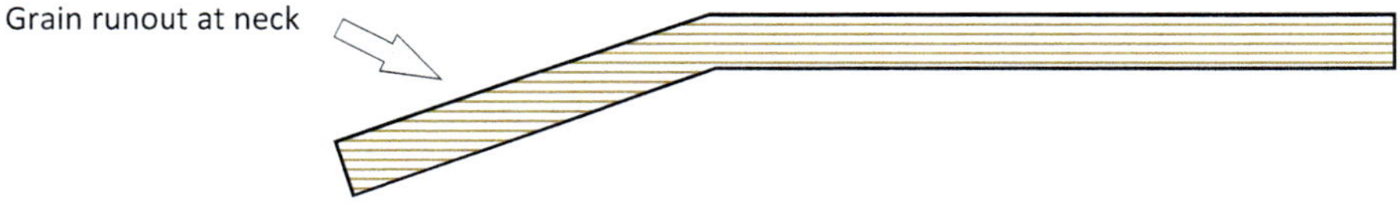

Fig. 2.53 A neck and headstock sawn from a single slab

Another traditional way to make an angled headstock is simply sawing the heel, neck, and headstock as a single piece from a slab of wood. This simplifies construction, though the grain on the headstock has serious runout (Fig. 2.53).

Some Lists from George Gruhn

George Gruhn has seen more of the guitar industry than most and offers these lists as a way of organizing one's thinking.

Types of Guitar Buyer:

1. Utility Tool User
2. Collector
3. Speculator

Four Reasons People Buy a Guitar:

1. Want to learn to play
2. Have learned to play and want a better one
3. Have skills but want to explore a different style of instrument or different style of music
4. Like collecting guitars

A Guitar must have Balance from String to String:

1. Balance in volume
2. Equal sustain
3. Balance in dynamic range
4. Balance in harmonic complexity
5. Balance in clarity of articulation

Three Ways to Compete in the Guitar Industry:

1. Price
2. Quality
3. Innovation

Four Major Concerns of the Guitar Designer:

1. Design
2. Construction
3. Materials
4. Cosmetics

While this is a potential failure point, fractures generally only result from rough handling or an instrument being dropped. For example, Martin still uses single piece necks on some of their models, apparently without undue problems. Figure 2.54 shows a pair of one-piece necks in the Martin Factory. The grain clearly extends from the headstocks to the necks.

George Gruhn, vintage guitar expert and owner of the landmark store in Nashville, notes:

> We see far too many guitars which have been damaged by being dropped or receiving an impact which breaks the peghead. A neck with a properly glued well-designed peghead graft is far stronger than a one-piece neck of similar dimensions.

Note that peghead is another common name for the headstock and peghead graft is another name for the headstock joint.

Fig. 2.54 One piece necks during production process (image courtesy of Jason Ahner, Archives and Museum Manager, C.F. Martin & Co, martinguitar.com)

2.5 Modern Steel String Guitars

There are many steel string guitar manufacturers and their designs range from very traditional to ones that are highly unique. Let's take a short look at the range of instruments available and the features they have in common, starting with the range of sizes now available.

Modern steel string guitars come in many different sizes, from instruments the size of tenor ukuleles to those nearly as large as short scale basses. Fig. 2.55 shows an Ibanez Piccolo EWP14 guitar. It is intended for travelers or guitarists who want a ukulele-sized instrument. It has the same scale length as a tenor ukulele (17″/432 mm). The body is 12″/305 mm long, 9.375″/238 mm wide at the lower bout and 2.75″/69.9 mm deep. Because of its small size, its tuning is five half steps higher than standard, ADGCEA rather than EADGBE. This is about as small as a practical guitar can be, particularly for players with large hands.

On the other end of the scale is the Taylor 316e Baritone 8 guitar shown in Fig. 2.56. It has 8 strings, though Taylor also offers baritone guitars with 6 strings. The scale length is 27 in./686 mm, the body is 20 in./508 mm long and

Fig. 2.55 A "Piccolo" guitar, Model EWP14 by Ibanez (image courtesy of Sweetwater Sound, sweetwater.com)

Fig. 2.56 A Taylor 316e Baritone 8 LTD guitar (image courtesy of Taylor Guitars, taylorguitars.com)

16.25 in./413 mm wide at the lower bout. This is large for a practical guitar, though a few jumbo guitars have 21 /533 mm long bodies and some archtop guitars have 17 in./432 mm and occasionally 18 in./457 mm lower bouts.

2.5.1 *Body Sizes*

Different sized bodies need some way to identify them, and manufacturers often give them proprietary names. However, some manufacturers and many luthiers follow the Martin naming convention for their guitars. Martin has been making guitars for so long and they are so familiar that some of the names are now generic, like referring to any disposable tissue as a Kleenex. A good example is the Dreadnought (also occasionally spelled Dreadnaught), introduced by Martin in the 1931 after making similar instruments for the Oliver Ditson Co., beginning in 1916. There is hardly an acoustic guitar manufacturer that doesn't produce a Dreadnought style instrument, though they sometimes have different names.

Some Martin body sizes were originally given numbers, but curiously in what would seem reverse order, so that smaller numbers corresponded to bigger guitars. Size 5 guitars were very small, with a scale length of 21.1 in. or 22 in. (543.6 mm or 558.8 mm) and a body length of 16 in. (406.4 mm). Figure 2.57 shows a modern Martin Size 5 instrument, also called a Terz guitar. This instrument is intended to be tuned three half steps, a minor third, above standard and Terz is the German word for third.

Martin introduced a succession of larger instruments, with decreasing size numbers, such as the Size 1 shown in Fig. 2.40. The next larger instrument was the Size O. Rather than going to negative numbers, the next larger instrument was called the OO. A still larger instrument was designated the OOO and there was even an OOOO, though it is not currently in production. Not all Martin guitars follow this system. For example, the dreadnought body shape is designated by D and the jumbo is designated by J.

Fig. 2.57 A Martin Size 5 Terz guitar compared to a full-sized instrument (image courtesy of C.F. Martin Archives, martinguitar.com)

Martin has manufactured an almost mystifying number of models, so some further explanation might help. The body size is generally designated by letters while a following number refers to a set of aesthetic design features, essentially trim levels. A higher number indicates more elaborate features. A trailing "m" indicates all mahogany construction. Thus, the OOO-15m currently in production is made of mahogany with no binding and a single line of purfling in place of a rosette.

One of the most famous Martin guitars is the OM-28. OM refers to "Orchestra Model," a body shape similar to their OOO body. An OM-28 has a 14-fret neck, a Sitka spruce top, and rosewood back and sides. The differences between related models in the Martin product line can be small and sometimes only real aficionados can readily identify specific models. Martin designations are just names and are not very descriptive. Still, they are widely enough known that many other manufacturers use them as well.

Table 2.2 lists dimensions of a representative group of Martin guitar designs. Dimensions of Martin guitars are almost always reported in fractional inches, so they are listed that way here. A more complete summary is available in Martin Guitars: A Technical Reference, the authoritative book by Johnston, Boak, and Longworth. Finally, note that Martin body dimensions are all over the internet and they sometimes disagree. Those presented here are from Johnston, Boak, and Longworth because of their close association with Martin. Dick Boak retired from a long career at Martin where, along with his formal responsibilities, he took on the role of company historian.

Fractional inches are increasingly antiquated and we must assume they will eventually be replaced by metric units. Table 2.3 lists the same dimensions in millimeters.

Table 2.2 Sizes of Martin guitars—fractional inches

Type	No. of frets	Total length	Body length	Upper bout width	Lower bout width	Depth at neck	Depth at tail	Scale length
O	12	37-3/4	19-1/8	9-1/2	13-1/2	3-3/8	4-3/16	24.9
OO	12	37-3/4	19-5/8	9-3/4	14-1/8	3-1/4	4-1/16	24.9[a]
OOO	12	39-9/16	20-7/16	10-3/4	15	3-1/4	4-1/16	25.4
D	12	39-9/16	20-15/16	11-1/2	15-5/8	3-15/16	4-3/4	25.4
O	14	38-3/8	18-3/8	10	13-1/2	3-13/32	4-1/4	24.9
OO	14	38-5/8	18-7/8	10-7/8	13-5/16	3-11/32	4-1/8	24.9
OO Deep Body	14	38-5/8	18-7/8	10-7/8	13-5/16	3-27/32	4-5/8	24.9
OOO	14	39-3/8	19-3/8	11-13/32	15-1/4	3-1/4	4-1/8	24.9
OM	14	39-1/2	19-3/8	11-13/32	15-1/4	3-1/4	4-1/8	25.4
OOOO/M	14	40-3/8	20-1/8	11-11/16	16	3-5/16	4-1/8	25.4
D	14	40-1/4	20	11-1/2	15-5/8	3-15/16	4-7/8	25.4
J	14	40-3/8	20-1/8	11-11/16	16	3-15/16	4-7/8	25.4

[a]Some 12 Fret OO models had a 25.4 in scale length

Table 2.3 Sizes of Martin guitars—millimeters

Type	No. of frets	Total length	Body length	Upper bout width	Lower bout width	Depth at neck	Depth at tail	Scale length
O	12	958.9	485.8	241.3	342.9	85.7	106.4	632.5
OO	12	958.9	498.5	247.7	358.8	82.6	103.2	632.5
OOO	12	1004.9	519.1	273.1	381.0	82.6	103.2	645.2
D	12	1004.9	531.8	292.1	396.9	100.0	120.7	645.2
O	14	974.7	466.7	254.0	342.9	86.5	108.0	632.5
OO	14	981.1	479.4	276.2	338.1	84.9	104.8	632.5
OO Deep Body	14	981.1	479.4	276.2	338.1	97.6	117.5	632.5
OOO	14	1000.1	492.1	289.7	387.4	82.6	104.8	632.5
OM	14	1003.3	492.1	289.7	387.4	82.6	104.8	645.2
OOOO/M	14	1025.5	511.2	296.9	406.4	84.1	104.8	645.2
D	14	1022.4	508.0	292.1	396.9	100	123.8	645.2
J	14	1025.5	511.2	296.9	406.4	100	123.8	645.2

An underreported dimension in acoustic guitars is interior volume. Robert Corwin (vintagemartin.com) lists the volume of a 14 fret OO as 800 in.^{3} (13,110 cm^3) and a 14 fret O as 754 in.^{3} (12,356 cm^3).

Taylor, the other big American manufacturer of acoustic guitars, uses a different naming convention for their guitars. Their body sizes have names and much of their line has a corresponding three-digit model number. For these instruments, the first digit designates a build level and the second two digits are the body shape. For example, an entry level guitar is identified as being in the 100 series. A Grand Auditorium body guitar (a little bigger than an OOO/OM) is body series 14. A trailing e indicates onboard electronics—one or more pickups and a preamp—and a c indicates cutaway. Thus, an entry level Grand Auditorium guitar with a pickup and preamp, but no cutaway is model 114e. A top of the line Grand Auditorium model with cutaway and electronics is model 914ce. The Taylor line is expanding and will likely include instruments not listed here (Table 2.4).

2.5.2 *Steel String Guitar Materials*

Modern steel string acoustic guitars show the effect of material developments over the last 50 years or so, but the differences are usually not dramatic. One thing hasn't changed from the earliest days—steel string acoustic guitars are still almost universally made from wood. Inexpensive instruments often use laminated wood (plywood) and more expensive ones are made from solid wood. But, even modestly priced instruments usually have solid wood tops. Tops are almost always some species of spruce. Sitka spruce is the most common, but other varieties are used. Western red cedar is more rarely used and redwood appears occasionally.

Table 2.4 Taylor body sizes and model numbers

Name	Model no.	Body length (in./mm)	Body width (in./mm)	Body depth (in./mm)	Scale length (in./mm)
Baby		15.75/400.1	12.5/317.5	3.375/85.7	22.75/577.9
Big Baby		19.5/495.3	15.1/383.5	4.0/101.6	25.5/647.7
GS Mini		17.625/447.7	14.375/365.1	4.438/112.7	23.5/596.9
Academy 12	A12	19.5/495.3	15.0/381	4.375/111.1	24.875/631.8
Academy 10	A10	20.0/508	16.0/406.4	4.625/117.5	24.875/631.8
Grand Theater (GT)	n11/n21	18.5/469.9	15.0/381	4.25/108	24.125/612.8
Grand Concert	n12/n22/n52/n62	19.5/495.3	15.0/381	4.375/111.1	24.875/631.8
Grand Auditorium	n14	20.0/508	16.0/406.4	4.625/117.5	25.5/647.7
Grand Pacific	n17	20.0/508	16.0/406.4	4.625/117.5	25.5/647.7
Grand Symphony	n16/n26	20.0/508	16.25/412.8	4.625/117.5	24.875/631.8
Dreadnought	n10	20.0/508	16.0/406.4	4.625/117.5	25.5/647.7
Grand Orchestra	n18	20.625/523.9	16.75/425.5	5.0/127	25.5/647.7
Jumbo	n15	21.0/533.4	17.0/431.8	4.625/117.5	25.5/647.7

In the early twentieth century, sides and backs had widely been made from Brazilian rosewood or mahogany. However Brazilian rosewood is now an endangered species that is rightly protected by international treaty. Fortunately, the materials used for the backs and sides have only a small effect on the sound of the instrument and manufacturers are exploring many species they had previously overlooked.

There are far too many to list here, but the modern luthier can now choose among dozens of species. Fortunately, the market has enthusiastically accepted this much wider range of woods. Apart from being more environmentally sustainable, it offers aesthetic choices that wouldn't have been possible earlier.

The market for acoustic guitars includes a small number of composite instruments. They are usually made from graphite or a proprietary composite, following practices originally developed for the aerospace industry. Perhaps the most famous brand built around composite instruments is Ovation, a company that spun off from Kaman, a helicopter manufacturer.

Their most familiar composite guitar is probably the Adamas, notably used by Glen Campbell and Neil Diamond. The more expensive Ovation instruments use carbon fiber while lower priced ones use a proprietary glass fiber and resin laminate they call Lyrachord. Their signature feature is a molded bowl back. While some of their higher end instruments have composite soundboards, the product line includes many instruments with wood tops. Figure 2.58 shows a 1988 Ovation Elite with a wood top and Ovation's distinctive pattern of soundholes.

Fig. 2.58 A 1988 Ovation Elite with a wood top and composite bowl back, front view (image courtesy of Mike and Mike's Guitar Bar, mmguitarbar.com)

Fig. 2.59 A 1988 Ovation Elite with a wood top and composite bowl back, back view (image courtesy of Mike and Mike's Guitar Bar, mmguitarbar.com)

Whatever the merits of this soundhole pattern, it offers almost no access to the inside of the guitar. Ovation used the design flexibility afforded by the composite back to include a round access port (Fig. 2.59).

Fig. 2.60 A RainSong Black Ice Series Carbon Fiber Guitar (image courtesy of RainSong, rainsong.com)

As I write this, there are a few other small manufacturers offering graphite instruments and it's not clear when or if they will grow to hold a significant share of the market. One of the most established manufacturers of all carbon fiber acoustic guitars is RainSong, which offers a range of body shapes and layups. Figure 2.60 shows an instrument with their WS body shape (slightly larger than an OM). The pieced carbon fiber laminations are visually interesting. An instrument I played sounded good and the carbon fiber construction should be both durable and very tolerant of humidity changes.

Composite materials are strong, durable, and stable, which makes them attractive for guitars. However, they are typically expensive, with the cheapest all-graphite guitars costing as much as a mid-level Martin or Taylor. Another possible barrier has been that graphite guitars often don't always sound like wood guitars—the instruments that players have come to love. This isn't necessarily bad, just different, but it may take time for the market to become comfortable with that sound. Of course, it is also possible that designers may find a way for graphite guitars to more closely mimic the sound of wood ones. Along these lines, RainSong offers a series of instrument called Concert Series with unidirectional carbon fiber tops intended to sound more like a wood guitar. They also offer tops with thin wood veneers to mimic the look of a traditional wood guitar.

A newer manufacturer of carbon fiber instruments is Emerald Guitars, an Irish company with an impressively wide range of designs. Figure 2.61 shows their popular X20 model. It is a six-string acoustic guitar made almost exclusively of carbon fiber. The only exception in this picture is a thin quilted maple veneer on the soundboard, an optional decorative feature with no tonal purpose. The offset soundhole is a trademark feature of Emerald guitars. Note also the arm bevel molded into the lower bout.

Fig. 2.61 An Emerald Guitars X20 with a quilted maple soundboard veneer (image courtesy of Emerald Guitars, emeraldguitars.com)

One historically interesting attempt at using an alternative material is the Maccaferri G40 (Fig. 2.62). Introduced in 1953, it was made from Dow Styron (polystyrene). It was intended as a serious instrument as opposed to inexpensive plastic toys. Though it was produced until 1964, it had no lasting effect on acoustic guitar design and is now obscure.

2.5.3 *Wood in Steel String Guitars*

Wood has two properties that are problematic. The first is that it is hygroscopic—it readily absorbs water from the air and expands or contracts in response to the humidity in the air. Within reasonable limits this is tolerable, though extended exposure to very wet or, even worse, very dry air can damage wood acoustic guitars.

Another, more insidious problem is that wood creeps under load. That means it permanently changes shape over time, even when not loaded to near its failure stress. Metals creep, but only at high temperature, and many polymers creep, including those used in some popular wood glues. When the wood in an acoustic guitar creeps under string loads, the alignment between the neck and body can change, requiring a neck reset.

It's also not uncommon for tops to slightly deform, with a slight cavity forming between the bridge and the soundhole and a slight dome forming behind the bridge. Creep takes time, often years, so instruments that appear to be structurally sound when new may seriously deform over time. In extreme cases, the instrument may

Fig. 2.62 Maccaferri G40 Guitar made from Polystyrene (image courtesy of Bernunzio Uptown Music, bernunzio.com)

even be structurally compromised. Figure 2.63 shows a lightly built guitar with serious deformation in the soundboard that happened over a period of years. Lightly built instruments can be responsive and sound very good. Designers may choose a lighter structure with the understanding that it may shorten the life of the instrument.

Wood does have disadvantages. However, it is durable enough (there are 100-year-old guitars still being played), inexpensive, and is easy to work with basic tools. It's also very familiar and has an organic appeal that is hard to match with composite instruments. It appears for now that most guitars will be made of wood for some time to come.

Sustainability is a problem and collective action is working to assure a future supply of guitar wood. Some suppliers, such as Alaska Specialty Woods, are

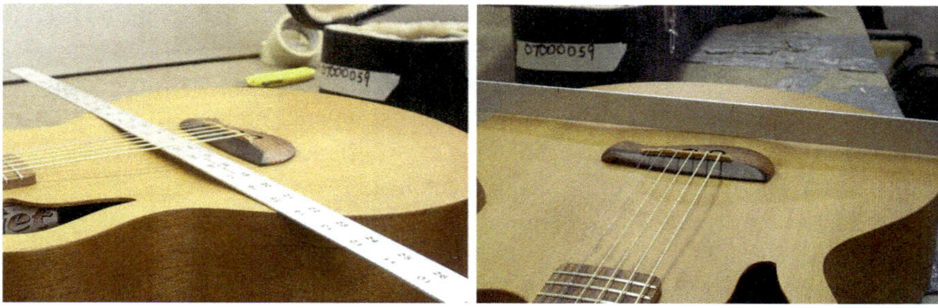

Fig. 2.63 An acoustic guitar with serious deformation in soundboard

specializing in salvaged and recovered wood and a few forward-looking people are replanting endangered woods. Bob Taylor, who along with Kurt Listug founded Taylor Guitars, is currently leading the way in replanting both koa and ebony. At this writing, some manufacturers are offering instruments made using "urban woods," trees grown in cities and cut as part of routine maintenance.

It seems that any group of luthiers, given enough coffee, tea, or beer, will eventually start talking about the importance of the quality of the wood they use. Opinions often run along a spectrum whose opposing ends are: a) Luthiers must always use the finest materials they can afford and b) A good luthier can make a good guitar from any materials that come to hand. These conversations have often revolved around shared empirical knowledge and personal experiences. Fortunately, enough formal test data now exists to nudge the conversation forward.

The clearest results are for back and side woods. As mentioned earlier, Torres famously made a successful classical guitar with cardboard back and sides. Modern luthiers have occasionally repeated the experiment and with the same result. Both historical experience and careful research point to the same conclusion: luthiers should feel free to make backs and sides from just about any wood they like.

Of course, this is not an absolute. There may be some very unusual or very impractical choice of woods that does indeed hurt sound quality—balsa sides with an ironwood back or something equally unlikely, perhaps. However, available data says that any practical choice of wood for back and sides should have a minor effect on sound quality.

A more difficult question is the importance of the quality of wood for the top. The most successful luthiers generally use the best grades of top wood and, unsurprisingly, they generally make very fine instruments.

The conversation must revolve around what we mean by high-quality wood. Luthiers mostly agree on a few traditional characteristics:

1. Fine wood has straight, even grain with closely spaced growth rings.
2. It is relatively stiff when flexed by hand across the grain.
3. It rings well when held lightly and tapped with a fingertip.

While these tests are perhaps a bit idiomatic, they don't often give the wrong answer; wood that passes these tests can reliably be used to make fine guitars. But, that doesn't necessarily mean that wood failing some of these tests can't be used to make good guitars. This is what statisticians call a false negative.

We have to assume that the generations of luthiers who came before us noticed the effect of wood quality on the sound of their instruments, and that the preferences handed down to us rest on a firm foundation of experience. Still, there may be some wiggle room, and the evidence seems to come from two sources.

The first is empirical. Many builders, either out of curiosity or necessity, have made guitars from inferior wood and some of these instruments have turned out well. A subset of these is "lumber yard guitars"—instruments made from whatever was available at the local lumber yard or home center. It seems that every luthier needs to make at least one of these just to get it out of their system (I made a couple). One of the most well-known is Bob Benedetto's Knotty Pine Guitar, shown in Fig. 2.64. He made the top from a very average 2″ × 10″ softwood plank from a

Fig. 2.64 The Benedetto "Knotty Pine Guitar" (image courtesy of Benedetto Guitars, benedettoguitars.com)

lumber yard. Wood for the back and sides came from his scrap bin. In discussing it, he says it sounds "fine," and I'm sure he's right. However, he also said that it may have sounded better if he'd used better wood. He's probably right about that, too.

The second is from published research. A 2019 article by Sebastian Merchel and Ercan Altinsoy from Dresden University of Technology and David Olsen, an otological surgeon working with Pacific Rim Tonewoods, describes acoustic test of a group of similar, purpose-built instruments. Taylor Guitars made 10 nominally identical model 814ce acoustic guitars with carefully selected woods. All but the top and brace wood were selected to be as identical as possible. Top and brace woods were selected based on stiffness and density and then the tops and braces were made to the same dimensions on all 10 instruments. Listeners preferred instruments made with lighter, more flexible top and brace wood, wood that may have been rejected based on traditional empirical testing.

This work is promising and may be influential, but we'll need more evidence to be sure. It suggests that it's possible to make high-quality instruments from what would traditionally be considered lower grade wood.

Factories generally make instruments to fixed dimensions, no matter what the mechanical properties of the wood are. While that could be a source of variation, dynamic testing on instruments at the Taylor factory has shown that the variation in the first few resonant frequencies is very small, with standard deviation of the first body resonant frequencies on the order of a few percent. Standard deviation is a statistical measure of how spread out data is. Low standard deviation means that the data tends to be close to an average value. This result doesn't mean that any wood at all can be used to make guitar tops. Rather, the variation was small in pool of guitars made from wood selected for Taylor by their supplier.

Glue is another important material. Steel string acoustic guitars are almost all assembled with yellow wood glue (aliphatic resin or AR). AR is attractive for several reasons—it is inexpensive, nontoxic, easy to apply, strong, and dries quickly. Some formulations even include an ultraviolet dye that glows under black light, so it's easy to find excess glue. There are also different formulations of AR for differing resistance to moisture. The most popular brand in the US right now is probably Titebond. Titebond original is the default glue for many luthiers. Titebond II and Titebond III are successively more resistant to moisture and about equally strong.

There are a few problems with AR. One is that it doesn't stick to itself like hide glue, so it is used only for joining bare wood surfaces. It also softens with heat and the most popular kinds can weaken in extremely wet conditions. It's very rare for a guitar to get wet enough for AR glue to fail and that would probably be catastrophic for other reasons as well. For example, the only catastrophic AR glue failure I have personally seen is an instrument that was immersed when the owner's house flooded.

If there is a problem with AR glue, it is that some formulations of it don't cure to a particularly hard condition. That means a joint glued with it could add damping to the structure. This is particularly a problem with poorly fitted joints where the glue layer is thick. The obvious solution is to ensure that joints are very tight, so the glue layer is thin. Trevor Gore and Gerard Gilet observe that it is nearly impossible to tell which glue is used in a well-made guitar just by listening to it. Glue is strongest in shear and in a thin, even layer. Thick glue layers are inherently weaker than thin ones as long as the joint is completely wetted and not starved for glue. There is no place for thick or uneven glue layers in acoustic guitars, no matter what glue is used.

Paco Chorobo is a Spanish luthier specializing in classical and flamenco guitars who also teaches guitar making. A former telematics engineer, he naturally explored the mechanical properties of glue as he refined his own methods. He states that wood being joined should fit "like glass on glass." In his class, he insists on glue layers thin enough that pencil marks can be read through the wet glue. In his own shop, in Úbeda, Spain, he uses one 16 oz. (473 ml) bottle of AR glue *per year*. He maintains that he has never had a glued joint fail, even in lightly built flamenco guitars (Fig. 2.65).

Fig. 2.65 Paco Chorobo applies glue to a fretboard while Robbie O'Brien supervises

2.5.4 *Steel String Guitar Structure*

Some structural elements of guitars are invisible to the players, but they have been refined since the Martin OM and Dreadnought became popular before the Second World War. Perhaps the biggest recent change to the structures of acoustic guitars is in neck joints, a feature invisible to the players. Many necks are bolted to the body and the traditional Martin glued dovetail is increasingly rare. Acoustic guitars currently use a variety of bolted neck joints. Figure 2.66 shows one offered by Stewart MacDonald, based on a common design.

A refined version of bolted neck joints was developed by Taylor Guitars. The most commonly used bolted neck joints make it easy to remove the neck, but changing the neck angle can require either very carefully removing material or improvising shims. A more intentional approach is to simply design shims into the neck joint so that they can be removed and replaced as necessary. Figure 2.67 shows a Taylor body on the production line with pockets for specially designed shims. The instrument cannot be assembled without them. During final assembly, the body and neck are aligned in a fixture and the correct thickness shims are inserted to maintain alignment. If the neck needs to be reset later, it is just unbolted, the old shims removed and new shims are inserted. Note also the small electromagnetic pickup between the soundhole and the top neck pocket.

The other important structural feature, the headstock joint, has probably changed less than the neck joint. Traditional scarf joints are very common and strong enough that they seldom fail, no matter what glue is used. When headstocks fail, it is often because of an impact. Less often, failures are caused by strings contracting in very cold environments, such as the cargo hold of an airliner.

Fig. 2.66 A bolt on neck and heel block offered by Stewart MacDonald (image courtesy of Stewart MacDonald, stewmac.com)

Fig. 2.67 The body of a Taylor Guitar with pockets for neck shims

Fig. 2.68 A finger joined headstock on a 2001 Taylor Guitar

In keeping with their ethos of continuous development, Taylor has used several different headstock joints. A marked departure from the traditional scarf joint was a finger joint, as shown in Fig. 2.68. This design lends itself to large-scale production since it can be made using a shaper—both sides of the joint are same and are just offset by half a "finger." It also offers a large gluing area that is in shear.

It's important to note that glue is most useful when connecting two parts that are loaded in shear. That is, they are trying to slide across one another. Thus, well-designed glued joints are mostly in shear.

Figure 2.69 shows a simple joint with the two glue layers in pure shear.

This is hardly a practical joint for guitars. However, a scarf joint with a reasonably low angle (10–15°) is closer than you might think to the ideal. A scarf joint

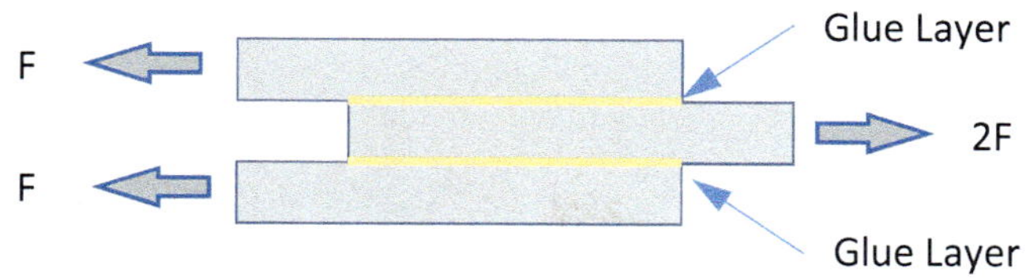

Fig. 2.69 An ideal glued joint with the glue in shear only

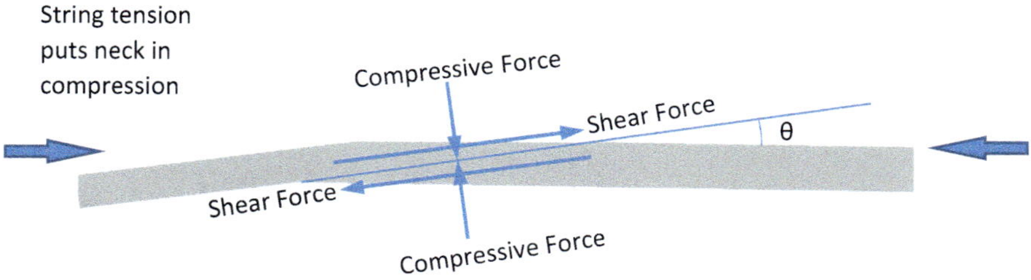

Fig. 2.70 Components of force in a scarf joint under compression

Fig. 2.71 Neck blanks with curved scarf joints in the Taylor Factory (image reproduced courtesy of Taylor Guitars, taylorguitars.com)

under string tension, is largely in shear, with a small compressive force component. Compression in a glue joint is much preferred over tension.

Figure 2.70 shows the components of force on a scarf joint under string tension. Most of the force in the joint is shear and, the lower the angle, the larger the portion in shear. When the angle of the glue joint, θ, approaches 0°, the portion of the string force transmitted in shear approaches 100%.

A more recent design from Taylor is a variation on the traditional scarf joint, shown in Fig. 2.71.

Fig. 2.72 A Taylor guitar headstock with a curved scarf joint (image courtesy of Taylor Guitars, taylorguitars.com)

The curved surface increases the gluing area that remains after shaping the neck and headstock, as shown in Fig. 2.72. This is a very strong joint and is unlikely to break, even under heavy use. Note that the glue seam intersects the back of the neck at a large enough angle that the seam is not accentuated.

Another important structural element in steel string acoustic guitars is the bracing pattern, and few design elements have been more thoroughly discussed. There are far too many different designs to catalog here, so let's focus instead on the big underlying idea. Top bracing does two important things. The first is that it strengthens and stiffens the top so it can better bear string loads. The second is that it acts as a mechanical filter to change the tone of the guitar.

A guitar top must first be strong enough to withstand string tension. The total tension of a set of guitar strings depends on tuning, scale length, and diameter. Thus, there is no single number for total string tension for a set of strings. However, it's enough here to use some representative numbers for full sized guitars (scale length = 25.5 in./647.7 mm) at standard tuning. Table 2.5 shows combined tensions for popular types of strings. The combined tension of a set of steel acoustic guitar strings is about the weight of an adult male. The combined tension of a set of steel strings is about double that of a set of classical strings, so the structure of a steel string guitar must be significantly stronger and heavier than that of a classical guitar.

Table 2.5 Typical total tensions of D'Addario string sets

Type	Material	Tension (lb)	Tension (kg)
Pro-Arté Normal Tension (EJ45)	Nylon	85.9	39.0
Pro-Arté Dynacore Hard Tension (EJ46TT)	Nylon	90.7	41.1
Silk and Steel (EJ40)	Steel	127	57.6
Phosphor Bronze Light (EJ16)	Steel	160.5	72.8
Phosphor Bronze Medium (EJ 17)	Steel	185.3	84.0
Phosphor Bronze Heavy (EJ18)	Steel	212.8	96.5
Phosphor Bronze Medium 12 String (EJ39)	Steel	320.0	145.1

Some readers will note that tension is given in kilograms (kg), though it is a unit of mass, not force. Alas, kg is popularly treated as a force everywhere but technical calculations, where forces are correctly expressed in Newtons (N).

Sitka spruce is both strong and light. The failure stress of Sitka spruce in compression is high enough (about 2300 psi or 15.9 MPa) that a small square stick 10 mm or about 3/8 in. on a side is theoretically enough to bear the loads from a heavy set of strings. For example, a set of D'Addario heavy phosphor bronze strings (EJ 18) has a combined tension of about 947 N (212.8 lb. or 96.5 kg) and our ideal square stick can bear 1590 N (357 lb. or 162 kg). However, the structure of an acoustic guitar top departs far from the ideal square stick. The top needs to distribute the string loads in a way that allows it to be flexible enough to do its job while being strong enough and durable enough to be practical.

Structures under load have a load path, essentially the path that forces take through the structure. The load path in a guitar top should ideally go straight down the center of the top, from the neck joint to the bridge. It's generally a bad idea to put a large hole in a structure in the middle of a load path, but most acoustic guitars do exactly this with the soundhole. Thus, the bracing must redirect the load path around the soundhole.

Engineering design almost always requires satisfying conflicting requirements. X Bracing is so deservedly popular because it solves the two conflicting problems of being strong enough to withstand string loads while still being flexible enough to respond to the vibration of those strings. Modern X-braced guitars often vary little from the original Martin design for the simple reason that it works so well. Figure 2.73 shows the top of a guitar made in the summer of 2018, but the bracing pattern would have looked familiar to C.F. Martin III 100 years earlier. A possible exception is the scalloped braces, which Martin introduced a bit later.

Still, luthiers are constantly experimenting with new solutions to old problems and many alternative bracing patterns have been developed. At this writing, a promising one is V bracing, developed through extended experiment by Andy Powers and Bob Taylor of Taylor Guitars and now in production. It solves the same problem that faced Martin 100 years ago and, they believe, does so with a subtle but valuable improvement in sound quality. Figure 2.74 shows the top of a Taylor guitar with the neat, simple V-brace pattern. Note the two-layer bridge patch and the routing around the lower bout. The routing reduces bending stiffness near the edge of lower bout, allowing the top to respond more readily to string vibration.

Fig. 2.73 An acoustic guitar top with X bracing

Fig. 2.74 The top of a Taylor guitar with V bracing (image courtesy of Taylor Guitars, taylorguitars.com)

One of the more dramatic departures from the traditional X-brace pattern is by Australian luthier and engineer, Trevor Gore. His design, which he calls Falcate bracing (Fig. 2.75), emerged from a careful study of the physics of the vibrating soundboard. Gore is one of the few luthiers with extensive technical training, which he has applied to the problem of designing acoustic guitars.

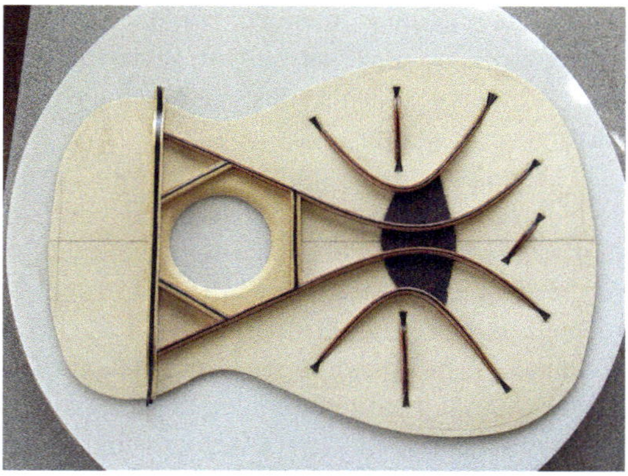

Fig. 2.75 Falcate bracing developed by Trevor Gore (image courtesy of Trevor Gore, goreguitars.com.au)

Trevor Gore on Guitar Design

Designing a guitar and building one are closely connected. It's all about having a comprehensive methodology that goes all the way from design through build, focusing on acoustics rather than just the woodwork.

I don't think designing and building should be separated. Rather, they are both important parts of a complete process. There are 3 main aspects I consider: playability, structural integrity and musicality. A fourth is visual aesthetic which I haven't written about much as it's such a divisive subject!

- Playability is all about instrument size and its set-up, neck shape, string spacing, ergonomics, etc. That's why I do wedge bodies, body shapes with high waists (easier to reach up the neck), tilt necks (user adjustable neck angle/action adjustment requiring no tools), arm rests and the like.
- Structural integrity is about making sure thin panels don't crack, guitars maintain their design geometry over a long period (use of composites) and are serviceable (bolt-on necks).
- Musicality is about ensuring an alluring sound (yes, even that can be defined!), getting the tone you want by understanding what modal resonances produce what types of sound, avoiding wolf tones by placing resonances in particular relationships and having the instrument play accurately in tune to the equally tempered scale using nut and saddle compensation.

The design methodology involves measuring the material properties of the wood you're about to use and a way to use those material properties so that top and back panels and their bracing resonate at or close to particular frequencies on the assembled guitar. Another comprehensive set of techniques, including the use of variable side mass are used to bring the guitar exactly to its design resonant frequencies.

Falcate bracing is intended to allow more precise adjustment of longitudinal and lateral stiffness since the two are more independent with this bracing pattern than with X bracing. It is also designed to make the most efficient use of material where the highest loading is.

These different bracing patterns still have to direct the load path around the soundhole at the center of the instrument. Of course, another approach to the problem of moving the load path around the soundhole is to just move the soundhole out of the way. Ovation did this with their multi-hole pattern and others have experimented as well. Figure 2.76 shows a MacPherson acoustic guitar with an offset soundhole. The soundhole is large enough that a repair person could reach a hand inside without having to remove the strings.

Another offset soundhole design is by Emerald Guitars, a company specializing in graphite instruments. Fig. 2.77 shows an Emerald X7 parlor guitar with their signature offset soundhole. Their product range currently includes 15 instruments, with all but one having offset soundholes.

Before moving on, it's important to mention curvature of top and back plates on acoustic guitars. While sometimes called flattop guitars, the soundboards and backs of these instruments are seldom truly flat. Builders learned long ago that building a slight dome into them made them stiffer, more durable, and allowed the top to be lighter and more responsive.

Traditionally, arching tops has been variable, with individual builders developing their own geometries and their own methods. The shape was often established by sanding arches into the braces and gluing them to the top. Spherical curvature is now common in steel string guitars, especially production ones.

Fig. 2.76 A McPherson acoustic guitar with an offset soundhole (image courtesy of Mike & Mike's Guitar Bar, mmguitarbar.com)

Fig. 2.77 An Emerald Model X7 with offset soundhole (image courtesy of Emerald Guitars, emeraldguitars.com)

Spherical curvature is just that—the top and back are sections from the surface of a sphere. There is no universally accepted radius, though tops are generally in the range of 25–40 ft. (7.62–12.2 m). Backs generally have smaller radii so they are more strongly curved. Back radius is usually in the range of 12–25 ft. (3.66–7.62 m). Figure 2.78 shows Charles Fox with an acoustic guitar whose top radius is 25 ft. The large straightedge is tangent to the top where the fretboard will overlap. A subtle flat has been sanded into the top to create a mating surface for the end of the fretboard (note the sanding dust near the edge of the top). The wide availability of computer controlled (CNC) routers means that individual builders can readily make or buy spherical dishes in which tops and backs are built.

A curved plate is stiffer in compression than an equivalent flat plate of the same material. To show the basic idea, try a simple experiment with an index card. Push on the ends of the card, putting it in compression and it buckles easily (Fig. 2.79).

Fig. 2.78 Charles Fox showing the curvature of a guitar top

Fig. 2.79 Buckling a flat index card

Fig. 2.80 A curved card in compression, resisting buckling

Now, rub the card across the hard edge of a table or desk to give it some permanent longitudinal curvature and push on the ends again. Even though it's the same card, it's now stiffer and able to resist more force than before (Fig. 2.80).

2.5.5 Steel String Guitar Finish

The finish on an acoustic guitar is there to protect the wood. Wood is strong, but relatively delicate and needs some type of coating to protect it from the dirt, water, and other gunk that are part of the life of guitars. Modern finishes can take many forms but are generally some type of polymer that is sprayed or brushed on. Traditional gloss finishes are finish sanded and buffed to a shine, though this is for appearance and isn't mechanically necessary. Satin finishes that don't require extensive buffing are common on acoustic guitars.

A useful way to organize finishes is by whether drying involves any chemistry. Some finishes dry by simple evaporation of a solvent and some undergo a chemical reaction as they cure. The two common finishes that dry by evaporation are shellac and lacquer. They are just solids dissolved in a solvent so that the cured finish on the instrument is still chemically identical to the original solid. Shellac is used as a sealer and a few luthiers apply it by French polishing, but lacquer is much more popular.

Nitrocellulose lacquer is partially nitrated cellulose (chemically related to guncotton) dissolved in a mixture of volatile solvents. Often called nitro lacquer, the modern version was developed in the 1920s by Edmund Flaherty at DuPont and marketed under the trade name Duco. When pigment was added, it was adopted by Ford as a quick-drying paint. Without pigment, it became a popular wood finish. Nitro lacquer was a step forward from the varnishes that had been used before and is still popular among individual and boutique builders.

It is easy to apply with either a brush or a spray gun. It is also clear and reasonably durable. Perhaps just as important is that it is easy to repair. Since there is no chemical reaction when it dries, subsequent layers partially dissolve the existing finish and "melt" into it. Figure 2.81 shows lacquer packaged in spray cans for small builders who don't have spraying equipment. A few cans, often including some with tints, is enough for a nice finish on a guitar.

However, nitro lacquer has some serious problems. The first is safety—the solvents in it are particularly noxious, so a cartridge breather and careful ventilation are essential. A second is that it takes a long time to fully dry. It dries to the touch relatively quickly, and instruments can often be handled after 4–8 h, but it can take weeks to dry hard enough to be buffed to a final finish. Guitar factories where lacquer is still used often have storage areas where instruments are allowed to finish curing. Lacquer also yellows and cracks with age, though it can be repaired.

As industrial chemistry has progressed, so too has the range of options available to guitar makers. Shellac, varnish, oil, and lacquer have largely been replaced in manufacturing operations. There are too many products on the market to list here, but manufacturers often use urethane or polyesters. These cure to hard, durable finishes that hold up well and both can be buffed to a high shine. Urethanes intended for buffing often include other resins. Modern finishes are often multi-part, in which

Fig. 2.81 Guitar lacquer packaged in spray cans (image courtesy of Stewart MacDonald, stewmac.com)

a catalyst is added to speed curing. Some catalysts cure by a chemical reaction and some cure by exposure to ultraviolet (UV) light which, in turn, catalyzes a chemical reaction. Catalyzed finishes require spray booths, breathing protection, and air filtration, so they are not typically found in smaller shops. However, at least one company is marketing catalyzed polyester in a two-component spray can. It still requires a breather, but can be simply sprayed outside if weather allows.

For small builders, a class of water-borne finishes has arrived on the market relatively recently (Fig. 2.82). They are sometimes called "water-based," but this is not strictly correct since typical formulations include some type of glycol. However, they clean up with water and emit no volatile organic compounds (zero VOC). They are also clear, unlike oils and oil-based polyurethanes that typically have an amber cast. Those formulated for guitars often dry a bit harder than general purpose versions sold in hardware and home improvement stores. Harder formulations sold as floor finish have worked well on guitars.

Any finish adds weight and heavy coats can affect the sound of the instrument. A more flexible finish can also add damping. Lightly damped structures vibrate longer

Fig. 2.82 Water-based polyurethane specially formulated for guitars (image courtesy of Stewart MacDonald, stewmac.com)

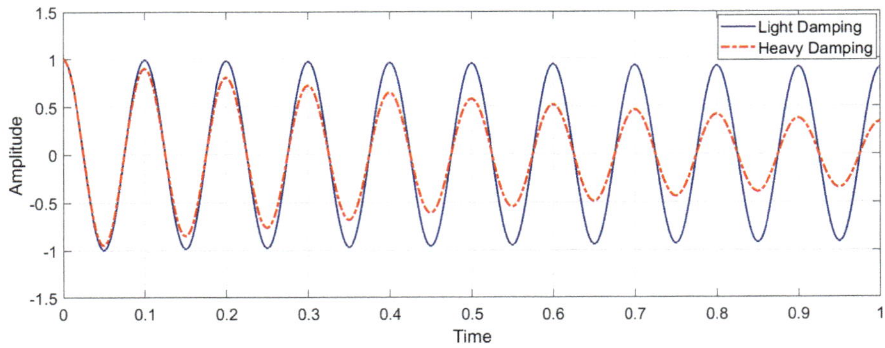

Fig. 2.83 Lightly damped and heavily damped vibration

than those with heavier damping. Figure 2.83 shows what happens when damping is added to a vibrating structure. The motion decays more quickly and the radiated sound dies off more quickly. Thus, an ideal finish is hard enough to add little damping.

Damping dissipates energy. In an acoustic guitar, the damping comes primarily from two sources, acoustic radiation and material damping. Acoustic radiation is just the mechanical energy from the vibrating structure being transferred to the air in the form of sound waves. The energy in the radiated sound is lost to the structure. Material damping is energy lost to the structure in a way that acts like friction and which turns some of the energy of the vibrating structure into heat. There isn't much mechanical energy in a vibrating guitar top, so the heating is far too small for the player to feel. The important point is that the finish can increase damping in the top, dissipating energy that might otherwise be radiated as sound.

The ideal guitar finish would be durable, easy to apply, and very thin so that it added little mass or damping. Getting any two of these is much easier than getting all three. For example, French polishing is thin and easy, if tedious, to apply. However, it is not very durable. Conversely, a coat of marine spar varnish is easy to apply and durable, but might be so heavy as to hurt the sound quality of the instrument.

One of the most successful industrial approaches to satisfying all three requirements is from Taylor Guitars. Some years ago, they started spraying guitars using industrial robots (Fig. 2.84), greatly improving control and repeatability. Their process uses a UV cure finish, applied precisely in a thin and uniform coat and immediately cured in a UV oven. The process is repeatable and surprisingly clean since there is little overspray. The finish is cured during a short exposure in a high-intensity UV oven, so there is no additional waiting time for finish to cure. They state that the matte body finish on their model AD17 (Grand Pacific body shape) is 2 mil thick (0.002 in. or 0.052 mm), a refinement made possible by a precise, consistent and highly automated finishing process.

Fig. 2.84 An automated spray booth at Taylor Guitars (photo by the author, permission courtesy of Taylor Guitars, taylorguitars.com)

Gloss finish needs to be buffed and this, too, is done with industrial robots. Figure 2.85 shows an enclosure in which an industrial robot is buffing guitars.

Engineers all know that build variation and build quality are inversely related—the more of one you have, the less of the other. The precision and repeatability of using robots for finishing helps reduce the variation in the completed guitars, so that it's possible to observe the effect of small design changes. This simple idea underlies quality manufacturing methods everywhere, so it's no surprise that it should work in guitar making as well.

2.5.6 Steel String Guitar Design

Acoustic guitar builders may fall into three basic schools of thought. The first are those who make copies of successful instruments. For example, some fine luthiers are happily making copies of Martin OM and dreadnought guitars, though with their

Fig. 2.85 Robotic guitar buffing enclosure at Taylor Guitars (photo by the author, permission courtesy of Taylor Guitars, taylorguitars.com)

own headstock design. The second is a group who have used a familiar design as a starting point and modified it to suit their own creative goals. The third is a small group who have tried to rethink the guitar based their own concepts. None of these approaches is better than the others and there is certainly room for all of us to practice our craft as we like. This reflected in the range of current designs.

Rosie Heydenrych, of Turnstone Guitars, is an example of a builder who has used familiar patterns as a starting point for her own designs. She is based in Surrey, England, and uses native woods in her instruments where it makes sense to do so. Figure 2.86 shows her carving the bracing on an X-braced back.

Figure 2.87 shows a Turnstone TG model, a grand auditorium sized instrument (slightly larger than an OM) that hints at its Martin heritage, but with refinements that distinguish it from the most traditional designs.

Fig. 2.86 Rosie Heydenrych carving X bracing on a back (image courtesy of Rosie Heydenrych, turnstoneguitars.co.uk)

Fig. 2.87 A TG Model from Turnstone Guitars (image courtesy of Rosie Heydenrych, turnstoneguitars.co.uk)

Rosie Heydenrych on Design

A design is where it all begins—it will be the blueprint that carries you forward to the completion of the product. It is your map when questions arise along the way, so it must therefore be thorough in detail.

Whilst aesthetics and lines in acoustic guitar design are of course important, the most vital aspect of your designs are the structural considerations. As well as ensuring a long life, with stability at the core of your decisions, it is also the beating heart of your instrument.

In this modern age of luthiery, it's also important to establish a unique visual identity for your instruments. This can be a hard balance to strike, between finding market acceptance while pushing the industry forward.

From a build perspective, a craftsperson must also think about their tools, the space within which they create and the skills they have to hand. A design is one thing to create, but to implement it into real life is another thing entirely.

My philosophy regarding design is that of understated elegance. I don't want to test or jar the eye with my shapes; they require fluidity of line and balance of weight. In turn, colour must complement and bring life to the instrument. Structural features must feel cohesive to the overall theme, encasing the result in a carefully considered final unity.

The road to learning this craft is long and winding. It's tough to reflect upon what I wish I had known when I started—my gut reaction is to dwell upon all of the mistakes I have made. However, I must remember that without them I wouldn't have arrived at the quality I maintain today.

Taking the third of these approaches can be a fun and instructive experiment, departing completely from familiar designs. Figure 2.88 shows such an experiment. This guitar was designed to be easy to hold and to use inexpensive materials. The top and back are from 1/8″ (3.2 mm) birch furniture plywood and most of the rest of the wood was salvaged from your author's scrap bin. It sounded remarkably like a guitar, with a nice bass, and was comfortable to play. It also stood upright on its own and could be safely leaned against a wall. It was modestly successful, interesting, and worth doing. It also suggested a new design direction. At the time, I hadn't seen anything else like it, but have since found that other luthiers had already explored body shapes like this one. There such long a history and so many luthiers working now that it's difficult to have a truly original idea.

A noteworthy member in this third group is Michihiro (Michi) Matsuda, whose Model M1 appears in Fig. 2.89. This instrument is unconventional by the standards of many luthiers, though it is clearly related to more conventional steel string acoustic guitars. The most obvious difference is a low D extension for the sixth string and a capo so that it can be quickly brought back to standard tuning. This instrument is conventional by Matsuda's standards and only hints at the range of his creative explorations. His deconstructed archtop that appears later is a more radical departure from tradition.

Fig. 2.88 Experimental guitar made by the author

Fig. 2.89 A Matsuda Model M1, No. 67 with low D extension and a built-in capo (Image courtesy of Michihiro Matsuda, matsudaguitars.com)

Michihiro Matsuda on Guitar Design

I used to think that following traditions and authorities is the way to make good guitars—for example, following 30's Martin guitars, or following masters' works, etc.

But, since there is no absolute "good sound" (at least, I haven't been able to find it so far), I started to think about how we define "good sound" and what "good sound" means.

Why do we tend to judge "good sound" only by audible sound? When you play guitars, you get a variety information from your all senses, not only your ears. When you process it in your mind, your knowledge, existing images in your mind, legacy, authority, etc., they all influence how you judge "good sound." Why don't we take these issues into consideration about "good sound"?

This leads me to another question. We often say that guitars are tools for players. But, is it true to say this?

What does "tool" mean in this context? Am I a tool maker? Is this what I like to do? I have been thinking about it, and I will continue to.

When I make guitars, I like to use the same creativities all "artists" use (musicians, painters, sculptors, performers, etc.). It would be great to express myself more and well in my guitar making. I am heading this direction.

At the same time, guitars have function. I understand that most of my customers want me to find new and better solutions to functional issues they have. It is like the discussion about the difference between art and design. I think that artistic expression can create art, but It is not always the source of good designs.

It is a very deep question for me. I don't have a good answer, yet but thinking about it gives me motivation.

Some unconventional design elements are for strictly practical reasons, to improve the tone or the flexibility of the instrument. One whose popularity seems to be growing is fan frets, also called a multiscale fretboard. This is an implementation of the idea patented by Ralph Novak in 1989, though the patent has since lapsed. The frets are not parallel, rather they are "fanned" so that the scale length increases from the treble to the bass strings. Figure 2.90 shows an acoustic guitar with fanned frets.

The multiscale neck is intended to improve the tone of the instrument and make it more comfortable to play. This system adds two new design parameters since the designer must choose both how far to increase the scale length for the low E string and which fret should be perpendicular to the centerline of the neck.

In guitar design, new ideas often have old origins. The idea of a multiscale instrument dates from at least 1620 and a type of multiscale fretboard was patented in 1900. The patent, awarded to E.A. Edgren (Patent 652,353), clearly shows a six

Fig. 2.90 A Turnstone TG acoustic guitar with fanned frets (image courtesy of Rosie Heydenrych, turnstoneguitars.co.uk)

stringed acoustic instrument with fanned frets and unusual inline tuners on the headstock (Fig. 2.91). While it predates modern fan fret guitars, this is an obscure patent and appears to have had little effect on instrument design.

2.6 Early Archtop Guitars

Archtop guitars are structurally distinct from classical and flattop steel string guitars. They evolved largely in the United States and reached their current form shortly before the Second World War. Credit for the development of the first archtop guitar usually goes to Orville Gibson, working in his shop in Kalamazoo MI in the 1890s. Gibson went on to produce a range of archtop instruments that are still in production by the company bearing his name. They have also influenced many other designs since their introduction.

A typical archtop guitar is structurally about halfway between a flattop guitar and a cello or other member of the violin family. Higher quality instruments traditionally have tops carved from solid wood, usually spruce, while cheaper instruments or those meant to be amplified often have laminated wood tops.

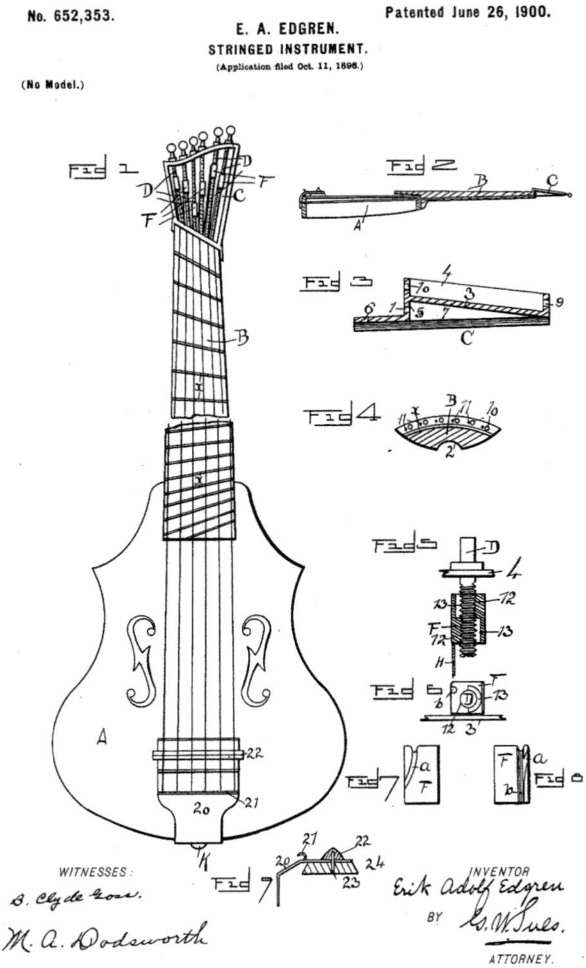

Fig. 2.91 An early patent for a stringed instrument with fan frets

Figure 2.92 shows a Gibson L-1 archtop from 1918. They reintroduced this model designation later as a flattop. Gibson was still using single round or oval soundholes. Later instruments generally used a pair of f-holes like those in the violin family.

Gibson introduced the L-5 in 1922 and it became one of their most successful models. Modern variations of it are still in production. One of its most influential players was Maybelle Carter, whose L-5, shown in Fig. 2.93, is probably familiar to anyone interested in the early days of country music. It was notably described by George Gruhn as "the most important single guitar in the history of country music." This instrument has most of the features now familiar in archtop acoustic guitars, including the carved solid spruce top, f-holes, a floating bridge held in place by string tension and a tailpiece fixed to the tail end of the guitar.

Fig. 2.92 A 1918 Gibson L-1 archtop guitar (image courtesy, Mike & Mike's Guitar Bar, mmguitarbar.com)

Fig. 2.93 A 1928 Gibson L-5 played by Maybelle Carter (image courtesy of Country Music Hall of Fame® and Museum, countrymusichalloffame.org)

Fig. 2.94 The Carter Family, including Maybelle (left) with her Gibson L-5 (image courtesy of Country Music Hall of Fame® and Museum, countrymusichalloffame.org)

The L-5 is a full-sized instrument, noticeably larger than a Martin OM. The lower bout was originally 16 in. wide (406.4 mm) and was increased to 17 in. (431.8 mm) in 1934. Carter's L-5 is one of the earlier 16 in models. Figure 2.94 shows the Carter Family: Maybelle, A.P. and Sara. It's clear that Maybelle, for her canonical presence in country music, was not very big. One wonders how she would have handled one of the larger, later versions. Also, one can't help noting that the instrument in the Country Music Hall of Fame has a different tailpiece than in Fig. 2.94. In the long working life of this instrument, the tailpiece must have been replaced.

While Maybelle Carter's L-5 is famous, archtops were more widely associated with other styles of music, particularly as flattops, especially dreadnoughts, grew to dominate country music. Archtops were some of the earliest guitars to be fitted with electromagnetic pickups, with Charlie Christian and Gibson leading the way. Gibson developed a series of archtop guitars with pickups and even the solid body Les Paul has an arched top clearly drawing on Gibson's famous archtops.

Archtop guitars most firmly established themselves in jazz. A familiar form of jazz archtop is a large acoustic guitar with a single cutaway and a floating pickup

Fig. 2.95 A new D'Angelico Excel and one from 1949 (image courtesy of D'Angelico Guitars, dangelicoguitars.com)

mounted to the end of the neck. It has a floating bridge, held to the top of the instrument only by string tension and a tailpiece that anchors the strings to the tail of the body. Figure 2.95 shows two similar D'Angelico archtop guitars that represent the classic New York jazz guitar. On the right is a restored 1949 Excel. On the left is a new Excel, a new design that draws on the older instrument. The new one has a floating pickup and an additional inlay on the neck between the 16th and 17th frets. John D'Angelico died in 1964. While the new D'Angelico company bears his name, it was started in 2011 and has no direct relationship to him.

Rock and roll demanded pickups, but hollow body guitars were susceptible to feedback. The result was that many archtops with pickups also had solid wood center sections, so they were not really hollow body guitars. The Gibson ES-335 may be the most well-known example of an electric archtop with what we now know as a semi-hollow body.

Truly hollow body archtops are now rare and ones without pickups are rarer still. Jazz players still often use archtop guitars with a pickup mounted on the end of the fretboard rather than directly to the top. Called a floating pickup, this arrangement allows the guitar to be amplified without adding the mass of a heavy electromagnetic pickup to the light top. The instrument on the left in Fig. 2.95 has a floating pickup.

2.7 Modern Archtop Guitars

Acoustic archtop guitars, including those with a floating pickup, are now a specialized type of instrument and a de facto design has emerged. The typical instrument is full sized with a body length of around 20in/508 mm and a lower bout width of 16–17 in. (406.4–431.8 mm). To accommodate the arched top, the neck is at an angle to the plane at which the top joins the body. This varies, but is often in the range of 4–5°, Bob Benedetto specifies an angle of 4.5° in his book on making archtop guitars.

Most instruments have cutaways for better access to higher frets. There are two popular types of cutaway, Venetian and Florentine. The Venetian cutaway is made from a single piece of wood and forms a smooth curve along the right side of the guitar.

Figure 2.96 shows a Benedetto La Venezia acoustic archtop with a Venetian cutaway.

Fig. 2.96 A Benedetto La Venezia with a Venetian cutaway (image courtesy of Bendetto Guitars, benedettoguitars.com)

Fig. 2.97 A Benedetto Andy Elite with a Florentine cutaway (image courtesy of Benedetto Guitars, benedettoguitars.com)

A Florentine cutaway is made from two pieces of wood that join to form a point. Figure 2.97 shows a Benedetto Andy Elite with a Florentine cutaway. This is a smaller guitar, with a 12 in./304.8 mm lower bout and a 23 in./584.2 mm scale length, and is marketed as a travel size instrument.

2.7.1 Archtop Guitar Body Sizes

Modern archtop guitars are routinely made in a range of body sizes, though they generally follow a traditional pattern, sorted by the width of the lower bout. The smallest full-sized instruments have a 15 in. (381 mm) width. The largest common instruments are 17 in. (431.8 mm) wide. A few instruments have 18 in. (457.2 mm) lower bouts, but this is about the practical upper limit. Few players are large enough to handle an instrument that big. Figure 2.98 shows a Benedetto Cremona with an 18 in lower bout.

Fig. 2.98 A Benedetto Cremona 18 (image courtesy of Benedetto Guitars, benedettoguitars.com)

2.7.2 Archtop Guitar Materials

At this writing, almost all archtop guitars are made of wood. Most tops are made of some type of spruce, with European spruce and Sitka spruce being popular. Sides and backs are often maple and often figured. However, just about any closed grain hardwood is structurally acceptable. The tops and backs are generally carved from solid blanks split from short sections of log, as shown in Fig. 2.99, though production guitars may have laminated tops or backs. The choice of back wood is conditioned by the need to carve it. Hard maple (Acer Saccharum) is durable and attractive when it is figured, but is difficult to carve. European white maple (Acer pseudoplatanus, also called Sycamore Maple) is easier to work using hand tools and is a popular choice for the backs and sides of archtop guitars.

There is at least one manufacturer offering an archtop guitar made completely of carbon fiber, the Kestrel by Emerald Guitars, shown in Fig. 2.100. While intriguing, at this writing it is still a rare instrument.

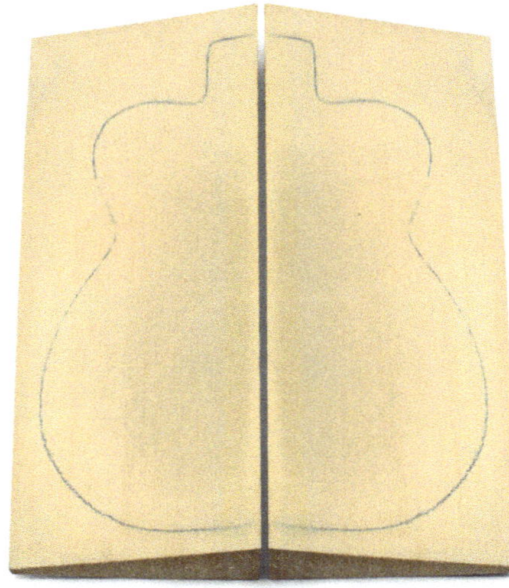

Fig. 2.99 An archtop guitar blank cut from European spruce (image courtesy of Stewart MacDonald, stewmac.com)

Fig. 2.100 A Kestrel carbon fiber archtop guitar by Emerald Guitars (image courtesy of Emerald Guitars, emeraldguitars.com)

2.7.3 Archtop Guitar Structure

Archtop acoustic guitars share some structural features with flattop, steel string acoustic guitars, but the arched top and back dictate the most important parts of the structure. One important structural feature is the floating bridge and its relationship to the arched top. It is held in place only by string tension and the strings are secured by a tailpiece that is, in turn, connected to an endpin, as shown in Fig. 2.101, or

Fig. 2.101 A 2003 Benedetto Manhattan with a tailpiece fastened to the endpin (image courtesy of Mike and Mike's Guitar Bar, mmguitarbar.com)

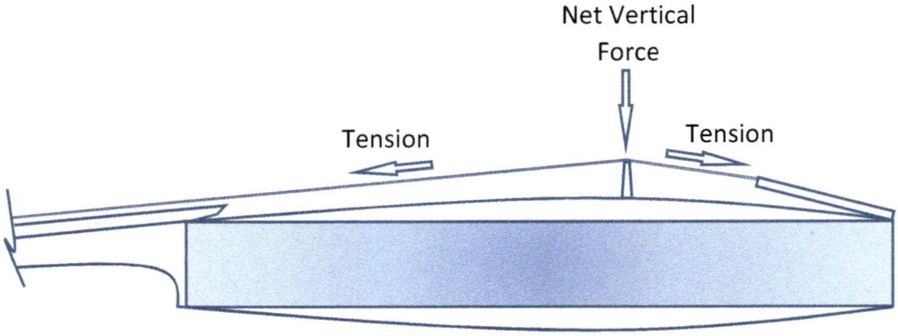

Fig. 2.102 String forces on a hollow body archtop guitar

simply screwed to the tail block of the instrument. This means that the bridge does not have to bear the string tension. Thus, the top doesn't need a bridge plate or any other structure to resist rotation of the bridge under string tension. Structurally, the top of a traditional archtop guitar is closer to that of a violin or cello than to a flat-top guitar.

Since the ends of the string are fixed to the headstock and the tailpiece, the body is in compression and there is no torque on the bridge as with a guitar whose strings are fixed to the bridge. There is a break angle over the bridge that creates a downward force, holding the bridge to the top. This floating bridge is held in place only with string tension and falls off if the strings are removed (Fig. 2.102).

Fig. 2.103 X Bracing on an arched guitar top by Bob Benedetto (image courtesy of Benedetto Guitars, benedettoguitars.com)

The structural mechanics of an arched wood guitar top are complicated and engineering descriptions involve phrases like orthotropic materials and doubly curved shells. Thus, the little experiment with a flat and then curved index card in Fig. 2.80 is only approximate. Still, perhaps it's enough to make the point that the elegantly curved top of an archtop guitar gives it stiffness properties different from flattops.

Archtops almost always have simpler bracing patterns than do flattop guitars. A large majority of archtop acoustic guitars use one of two bracing patterns, X brace or parallel brace. The X brace is much simpler than typically found on flattop guitars and usually consists of two crossed braces, as shown in Fig. 2.103. The braces are placed so that they directly support the feet of the bridge.

The other common type of bracing is called parallel bracing. The two braces aren't exactly parallel, but like many other things in guitar making, are close enough. Figure 2.104 shows Bob Benedetto trimming parallel braces on an arched guitar top. Arched tops are generally much thicker than flattops, with center thicknesses of 0.25 in. (6.35 mm) being common. A plan of a 1948 D'Angelico New Yorker, drawn by Steven Anderson and distributed by the Guild of American Luthiers, shows a top thickness of 0.310 in. (7.9 mm) at the center.

Fig. 2.104 Bob Benedetto trimming parallel braces on an arched guitar top (image courtesy of Benedetto Guitars, benedettoguitars.com)

2.7.4 Archtop Guitar Finish

Archtop guitars are generally finished with the same processes as flattop guitars. Given the strong influence of tradition in archtop design, nitro lacquer is still a popular finish. The finish on archtop guitars solves the same problem as it does on flattop steel string guitars, so it's no surprise that the finishes are similar.

2.7.5 Archtop Guitar Design

The archtop guitar is a refined instrument and many builders stick close to traditional patterns for their instruments. However, it's no surprise that some luthiers are experimenting with their designs. These experiments range from subtle and elegant refinements to completely new concepts of what an archtop guitar can be.

Maegen Wells has started with traditional designs, but is taking them in her own directions (Fig. 2.105). Apart from her own aesthetic touches, she is interested in expanding the musical range of archtop guitars and in instruments with smaller bodies.

Fig. 2.105 An archtop guitar with koa sides and back (image courtesy of Maegen Wells, maegen-wellsguitars.com)

Maegen Wells on Archtop Design

The designs of my instruments come from the foundation of traditional builders such as D'Angelico, D'Aquisto and, most significantly Tom Ribbecke, my mentor of many years—I had been bitten by the archtop bug before I started my 5-year apprenticeship with him, but I certainly wouldn't be where I am today without him. Above everyone and anyone else, he is at the foundation of everything I do.

From Tom's shop, I've gone on to specialize in small body acoustic archtops—I don't believe we have to have an 18" body to have a loud acoustic voice, and my archtops, as well as other builders out there, are proof of that concept. I mainly build 15" and 16″ archtops, and my mission is to help guitar players understand that contrary to popular belief, archtops are some of the most dynamic guitars in the world. You do not have to be a jazz player to play an archtop. I've often said that I feel like archtops have been put into a musical straight jacket. It is my goal to try and free the archtop from this limited reputation, by building archtops that excel in all genres of music and that are truly the best of both worlds between the acoustic and electric guitar.

While her work clearly shows her dedication to her craft, she appears to be unique among luthiers in having the design of an archtop guitar tattooed on her person (Fig. 2.106). This, dear readers, is commitment.

Fig. 2.106 Maegen Wells showing an elevated dedication to archtop guitars

Fig. 2.107 Benedetto Sinfonietta (image courtesy of Benedetto Guitars, www.benedetoguitars.com)

One popular area of design experimentation in archtop guitars is the soundholes, which are traditionally a pair of f-holes quite similar to those in the violin family. The soundhole in an acoustic guitar can significantly affect its tone. The primary effect of the soundhole is due to its area. However, the shape of the soundhole or soundholes has a secondary effect on tone. The f-holes that are so common in archtop guitars are sometimes replaced with other shapes. Figure 2.107 shows a Benedetto Sinfonietta with an oval soundhole, akin to those found on early Gibson archtops.

Bob Benedetto on Archtop Design

I'm sure I never made a guitar without consciously focusing on both acoustic and aesthetic designs, the latter of course being my personal preference and steeped in tradition.

Acoustic designs, which often influence aesthetics, begin with tonewood selection. Arguably, alternative woods have their place in modern guitar construction. However, on archtop guitars, my preference is lightweight, stiff conifers for the top and a lightweight, stiff variety of maple for the back, sides and neck.

Traditional parallel or X bracing patterns are preferred, although endless variations can successfully be incorporated into the overall acoustic design of any instrument. F-hole or soundhole sizes and locations have a dramatic effect on the voice of the instrument and must be given serious consideration throughout the design process.

Additional design appointments include body shape, width and depth; chemistry and thickness of the finish, even the weight of the tuning machines. Material selection and design of the fittings, which include the finger rest, tailpiece and bridge, are best made of wood (preferably ebony), which has an acoustical advantage over metals and plastics. Even string selection has an impact on the outcome of the instrument's voice.

Often, subtleties in design occur during my construction process. For instance, a particular carved top may produce an unexpected tone, which would affect the size of the F-holes. That, in turn, would affect the placement of the braces. The outcome would be slight variations to my original designs.

A more artistic take on soundhole geometry is shown in the Benedetto 35th Anniversary model (Fig. 2.108). This organic pattern serves the same purpose as f-holes or a round soundhole, but adds a nice aesthetic element.

A truly unique take on soundhole shapes is the Benedetto Il Teredo shown in Fig. 2.109. For this instrument, he chose a piece of Sitka spruce that had been used as part of a float. Over time, it was bored through by Teredo mollusks (shipworms). Bob salvaged this otherwise high-grade top and used the natural holes as the soundholes for the instrument. Since the holes are naturally occurring, no two instruments made this way would be alike.

Ken Parker is well known for innovative designs such as the Parker Fly electric guitar, but is also an archtop builder. His vision of the archtop guitar draws on familiar design elements, but moves well beyond more traditional designs. For example, the offset soundhole (Fig. 2.110) is intended to project sound both to the audience and the player. The headstock is similar to that used on the Parker Fly and includes carbon fiber reinforcement.

Fig. 2.108 35th Anniversary Benedetto Archtop (image courtesy of Benedetto Guitars, benedettoguitars.com)

Fig. 2.109 Benedetto Il Teredo with a worm-eaten top (image courtesy of Benedetto Guitars, www.benedettoguitars.com)

Fig. 2.110 An innovative archtop guitar by Ken Parker (image courtesy of Ken Parker, kenparker-archtops.com)

Less obvious is the way Parker has attached the neck to the body. Rather than the traditional neck joint, he has cantilevered the neck from a rectangular carbon fiber post set into the body (Fig. 2.111). Parker names his guitars and this one is named Mira.

A fluid, sculptural variation on archtop design is the Infinitum by Glenn Maxwell, a luthier in New Zealand. The soundholes are sculpted into the arched top and back and the strings are fixed directly to the top (Figs. 2.112 and 2.113). Like some other archtop makers, Maxwell uses a CNC router to rough out the tops and backs before refining the shapes by hand.

Fig. 2.111 Cantilevered neck joint and offset soundhole on an archtop guitar (image courtesy of Ken Parker, kenparkerarchtops.com)

Fig. 2.112 An archtop guitar by Glenn Maxwell—front view (image courtesy of Glenn Maxwell, maxwellguitars.com)

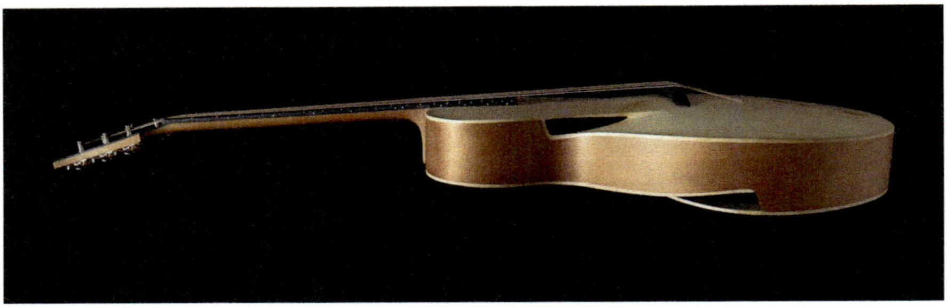

Fig. 2.113 An archtop guitar by Glenn Maxwell—side view (image courtesy of Glenn Maxwell, maxwellguitars.com)

Fig. 2.114 An archtop with cross-grain top (image courtesy of Glenn Maxwell, maxwellguitars.com)

An interesting experimental instrument is one Maxwell made with his signature geometry and a cross-grain Douglas fir top (Fig. 2.114). He wanted to show that the top bracing was able to resist string tension with less contribution from the top. While unconventional, even for him, it is a successful instrument that found a customer. He describes the tone as "surprisingly woody."

Fig. 2.115 Deconstructed Archtop guitar by Michihiro Matsuda (image courtesy of Michihiro Matsuda, matsudaguitars.com)

Before leaving modern archtop design, let's go to the extreme with another instrument by Michihiro Matsuda (Fig. 2.115). He calls it his Deconstructed Archtop and it is as much sculpture as guitar. It takes the concept of the archtop very far indeed from the traditional, though Matsuda is careful to ensure that his instruments, whatever their aesthetic statements, must first be fine guitars.

2.8 Hybrid Instruments

There is a class of guitars that occupy a middle ground between acoustic and electric guitars and includes the Gibson SST, the Taylor T5, and the Fender Acoustasonic series. They are intended to offer some flexibility missing from purely acoustic guitars, particularly when played on stage. They are less susceptible to feedback and have a combination of pickups that offer the player a wider range of tonal choices.

These hybrid instruments have thin bodies, usually milled from solid blanks, with cavities so that they are at least approximately hollow body instruments. They are fitted with acoustic bridges with fixed saddles. While they have some of the tonal character of acoustic instruments, they are only intended to be played through an amplifier.

Figure 2.116 shows a Taylor T5z, a slightly smaller development of the original T5. In addition to the electromagnetic bridge pickup that is visible in the picture, there is a soundboard pickup and another electromagnetic pickup in the neck. The neck is bolted on as with other Taylor instruments.

Since the back has little role in the dynamics, it has a removeable cover that allows access to the inside of the soundboard. The overall look of the instrument suggests a familiar Taylor acoustic guitar. Figure 2.117 shows a T5 body being bound in the Taylor Factory. While the edge binding is applied in the conventional way, the binding in the soundholes and the pickup pocket are molded as single pieces that fit tightly into the CNC milled pockets.

Tim Shaw on Guitar Design
Guitar design is as much about Context as it is about Concept. Here's what I've learned to ask as we start a project:

Why do we need another guitar? Who's the player or customer? What does this new one do that nothing else will? We need to have a vision (our Concept)

Fig. 2.116 A Taylor T5z (image courtesy of Taylor Guitars, taylorguitars.com)

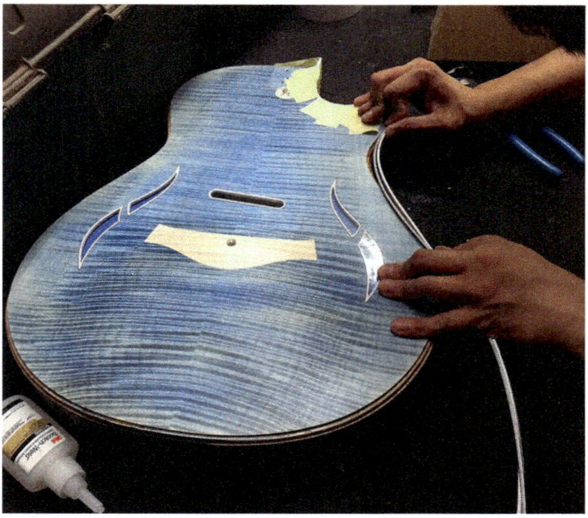

Fig. 2.117 A T5 being bound at the Taylor Factory

but there are millions of guitars out there and we're usually not designing in a world that's eagerly waiting to see what we do next.

Do we have the skills to make at least one of these? Now we're getting into the Context part of the equation. This is a loaded enough question for an individual builder, but on the factory level, it means knowing what the factory can do and whether you can produce a design correctly, repeatably, and at a profit.

This is hard enough if you work in the factory; it's much more difficult (as I learned the hard way, over and over) if the factory is on the other side of the planet.

Having a cool idea and building it to see what happens is the fun part of guitars. Then the Rest of the World steps in to complicate your life.

A more recent hybrid instrument approaches the design from the starting point of a solid body electric guitar rather than an acoustic guitar. The Fender Acoustasonic Telecaster (Fig. 2.118) has the features of hybrid instruments, but has successfully packaged them in the body shape of a Telecaster solid body electric guitar. It has more radiused edges and is contoured for improved comfort. A follow-on model uses the Stratocaster body shape.

One of the more interesting features of the instrument is the soundhole insert (Fig. 2.119). It was carefully designed to tune the air resonance of the body to create a peak in the response of the instrument. The result is that it has much better tone when unplugged than one might expect of a guitar with so little interior volume. While it could not be played on stage unplugged, it is possible to practice with it that way.

Fig. 2.118 An Acoustasonic Telecaster (image courtesy of Fender Music, fender.com)

Fig. 2.119 Soundhole insert in an Acoustasonic Stratocaster (image courtesy of Fender Music, fender.com)

Bibliography

Books

Benedetto, Robert, Making and Archtop Guitar 2nd ed., Centerstream Publishing 2018, ISBN 978-1574243550

Dobney, Jayson Kerr, Guitar Heroes: Legendary Craftsmen from Italy to New York, Metropolitan Museum of Art, 2011, ISBN 978-0300169805

French, Richard Mark, Technology of the Guitar, Springer 2012, ISBN 978-1461419204

Gore, Trevor and Gilet, Gerard, Contemporary Acoustic Guitar Design and Build, 2nd edition, Published by Trevor Gore, 2016, ISBN Vol 1: 978-0-9871174-2-7, Vol. 2: 978-0-9871174-3-4

Johnston, Richard, Boak, Dick and Longworth, Mike, Martin Guitars: A History, Hal Leonard Publishing, 2008, ISBN 978-0-634-03785-6

Johnston, Richard, Boak, Dick and Longworth, Mike, Martin Guitars: A Technical Reference, Hal Leonard Publishing, 2009, ISBN 978-1-4234-3982-0
Lawergren, Bo, The original of Musical Instruments and Sounds, Anthropos, Vol 83, 1988, pp. 31-45, www.jstor.org/stable/40461485
Morrish, John, The Classical Guitar Book, Backbeat Books, 2002, ISBN 978-0879307257
Mottola, R.M., Mottola's Cyclopedic Dictionary of Lutherie Terms, R.M. Mottola ISBN 978-1-7341256-0-3
Pensa, Rudy, Archtop Guitars: The Journey from Cremona to New York, Graphis, 2011, ISBN 978-1932026702
Romanillos, Jose L., Antonio de Torres Guitar Maker – His Life and Work, The Bold Strummer, 1997, ISBN 978-0906540855
Rudawska, Anna, ed., Adhesives: Applications and Properties, IntechOpen, 2016, ISBN 978-953-51-4148-8, DOI: https://doi.org/10.5772/62603. Free for download at: https://www.intechopen.com/books/adhesives-applications-and-properties
Sloane, Irving, Classic Guitar Construction, The Bold Strummer, 1989, ISBN 978-0933224148
Solondz, Simone, ed., Custom Guitars – A Complete Guide to Contemporary Handcrafted Guitars, String Letter Publishing, 2000, ISBN 978-1890490294
Somogyi, Ervin, Making the Responsive Guitar, Luthiers Press, 2009a, ISBN 978-0-9823207-1-6
Somogyi, Ervin, The Responsive Guitar, Luthiers Press, 2009b, ISBN 978-0-9823207-0-9
Turnbull, Harvey, The Guitar from the Renaissance to the Present Day, The Bold Strummer 1991, ISBN 978-0933224575
Urlik, Sheldon, A Collection of Fine Spanish Guitars from Torres to the Present, 2nd edition, Sunny Knoll Publishing, 2015, ISBN 978-0692467831
Westbrook, et al, Inventing the American Guitar: The Pre-Civil War Innovations of C.F. Martin and His Contemporaries, Hal Leonard Publishing, 2013, ISBN 978-1458405760

Articles

Anderson, Steven, 1948 D'Angelico New Yorker, American Lutherie #16, 1988. Also, Big Red Book of American Lutherie 2, Guild of American Luthiers, ISBN 0-9626447-3-0.
Bruné, R.E., Bridging the Gap, Guitar with Guts Series, Vintage Guitar, February 2010a, p. 78.
Bruné, R.E., A Baroque Guitar Restoration, American Lutherie #19, 1989. Also, Big Red Book of American Lutherie 2, Guild of American Luthiers, ISBN 0-9626447-3-0.
Bruné, R.E., What Wood Would Woody Wield?, Guitar with Guts Series, Vintage Guitar, September 2010b, p. 82.
Bruné, R.E., Spruce and Prejudice, Guitar with Guts Series, Vintage Guitar, March 2012, p. 100.
Bruné, R.E., Antonio de Torres – The Birth of the Modern Guitars, Vintage Guitar, September 2010c, p. 82.
Bruné, R.E., Andre Segovia's Hauser Built in 1937, Vintage Guitar, October 2003.
Carcagno, Samuele, et al, Effect of Back Wood Choice on the Perceived Quality of Steel-String Acoustic Guitars, Journal of the Acoustical Society of America, Vol 144, 2018, DOI: 1121/1.508475
Decker, John A. Jr., Guitar Acoustics 101, Guitarmaker #71, Spring 2010
Elliott, Jeffrey R., Open Harmonic Bar Classical Guitar, American Lutherie #127, P. 34, 2016
Elliott, Jeffrey R., 1930 Santos Hernandez Guitar with Segovia's Autograph, American Lutherie #115, P. 30, 2013a
Elliott, Jeffrey R., GAL Instrument Plan #68: 1930 Santos Hernandez Classical Guitar, American Lutherie #115, P. 36, 2013b
French, Richard Mark, Response Variation on a Group of Acoustic Guitars, Sound and Vibration, January 2008.

Gore, Trevor, Wood for Guitars, Acoustical Society of America, Proceedings of Meetings on Acoustics, Vol 12, 2011, DOI: https://doi.org/10.1121/1.3610500

Greven, John and Freeborn, Charles, Recreating the Vintage Martin Peghead Joint, American Lutherie, #102, P. 28, 2010, Also Flattop Guitars, Guild of American Lutherie, p. 50

Grimes, Steve, Archtop Guitars: Is Bigger Better?, American Lutherie #14, p.18, 1988. Also, Big Red Book of American Lutherie, Vol. 2, Guild of American Luthiers, ISBN 0-9626447-3-0.

Gruhn, George and Staff, Ditson's Style 11 and the Birth of the Dreadnought – Martin's Big Step, Vintage Guitar, July 2017.

Gruhn, George and Spann, Joe, The Real Story of Martin's 14 Fret Orchestra Models – OM Irony, Vintage Guitar, March 2018.

Gruhn, George, Martin D-28 and D-45 – Definitive Flat Tops, Vintage Guitar, April 2000a.

Gruhn, George, D'Angelico New Yorker – Top of the Line from a Master Builder, Vintage Guitar, March 2000b.

Johnston, R., The OM Style Guitar: What's in a Name and Does it Matter?, Guitarmaker #49, Summer 2004

Kamke, F.A. and Lee, J.N., Adhesive Penetration in Wood – a Review, Wood and Fiber Science, Vol. 39, No. 2, 2007, pp. 205-220.

Marchione, Stephen, Archtop Guitars with Italian Flare, American Lutherie #115, p.6, 2013

Manzer, Linda, et al, The Zeidler Guitar Project, Guitarmaker #46, Fall 2003

Meeker, W., Mark Erlewine's 40 Years with Willie Nelson's Trigger, Vintage Guitar, November 2017

Merchel, S., Altinsoy, M. E., Olson, D., Perceptual Evaluation of Bracewood and Soundboard Wood Variations on the Preference of a Steel String Acoustic Guitar, Journal of the Acoustical Society of America, Vol. 146, No. 4, October 2019, pp. 2608-2618.

Monteleone, John, A Modern Builder Interprets the Archtop, American Lutherie #6, p.34, 1986. Also, Also, Big Red Book of American Lutherie, Vol. 1, Guild of American Luthiers, ISBN 0-9626447-2-2.

Mottola, R.M., Measuring Scale Length of Fretted Instruments, American Lutherie #136, p. 69, 2019

Novak, Ralph, The Multi-Scale Fretboard, American Lutherie #21, 1990. Also, Big Red Book of American Lutherie, Vol. 2, Guild of American Luthiers, ISBN 0-9626447-3-0.

Olsen, Tim and D'Aquisto, James L., Building the Archtop Guitar, A Brief Overview, American Luthier #37, p.6, 1994a. Also, Big Red Book of American Lutherie, Vol. 4, Guild of American Luthiers, ISBN 978-0-9626447-6-4.

Olsen, Tim and D'Aquisto, James L., Building the Archtop Guitar, The Soundboard, American Luthier #37, p.10, 1994b. Also, Big Red Book of American Lutherie, Vol. 4, Guild of American Luthiers, ISBN 978-0-9626447-6-4.

Chapter 3
Iconic Guitars

The best luthiers understand some of the history of the guitar. Most have studied the precursors of their instruments and are familiar with the work of those who came before them. This understanding tends to be compact and practical, but can sometimes be quite broad, and occasionally nearly encyclopedic.

It makes sense for us to have a firm grounding in a few historically important instruments before designing new ones. A guitar designer with no historical perspective may make avoidable mistakes or waste time re-inventing old design elements. For context, I "discovered" the bridge ground for electric guitars and the Wheatstone bridge for resistors on my own before finding that both had been in common use for some time.

Historical understanding should be informative, but not restrictive. For example, having a clear understanding of the details and importance of Torres' 1856 guitar should be a foundation for exploring new designs, but not consign builders to a lifetime of making copies of it unless making faithful copies is their wish.

There have been far too many important designs over the last 150 years or so to list here. Rather than getting lost in that wilderness, let's focus on three specific designs, one each of a classical guitar, a flattop steel string guitar, and an archtop guitar. They are: Segovia's 1937 Hauser, a 1933 Martin OM-28, and a 1948 D'Angelico New Yorker.

I chose these three after much reading and after talking with several working luthiers. They are certainly not the only possible choices and I offer my apologies if your favorite instrument isn't among them. I chose them because they were highly influential instruments that are also well documented. Apart from influencing so many builders, detailed plans exist for all three of them, so modern builders can study them and even make replicas if they like.

R. M. French, *Acoustic Guitar Design*,
https://doi.org/10.1007/978-3-030-89381-1_3

3.1 1937 Hauser Classical Guitar Owned by Andrés Segovia

Even in the lutherie world, some guitars are as well known for their owners as for their builders. The classical guitar made in 1937 by Herman Hauser Sr. and owned by Andrés Segovia, the virtuoso Spanish guitarist, is certainly one of these. He played this instrument extensively in concerts and recordings. Segovia, who was famously picky about his guitars, famously described it as the "greatest guitar of our epoch." The stories of Hauser, Segovia, and how they met are well documented elsewhere. It is enough here to note that Segovia is generally credited with establishing the classical guitar as a legitimate concert instrument in Europe and the United States during the twentieth century. The classical guitar had a much higher status in South America and needed no additional push there.

Segovia's 1937 Hauser is only one of many instruments he played and he eventually set it aside in favor of others. However, it is probably the one most closely associated with him. One can hardly claim any knowledge of classical guitars without being familiar with this instrument. It has probably influenced modern luthiers as much as any other classical guitar.

This instrument has been the subject of numerous articles and some fine luthiers make copies of it. There are several plans of it in circulation. However, an accurate and particularly well-informed one is by Richard Bruné (GAL Instrument Plan #33-1937 Hermann Hauser, Sr. Classic, Ex Segovia), who knew Segovia personally and counts him among buyers of his own guitars. Bruné drew his plan after several days spent carefully measuring the original at the Metropolitan Museum of Art, where it is now in their permanent collection after having been donated by Segovia's widow, Emilita, in 1986. Figure 3.1 shows the neck and soundboard from Bruné's plan. It uses the familiar symmetric seven bar fan bracing pattern with two V braces, often called cutoff bars, pointing to the tail block, and three transverse harmonic bars.

In spite of its legendary status, one can't help noticing how unremarkable the design appears to be at first glance. Indeed, it appears just a standard classical guitar, almost generic in its familiarity. Only details of the design, albeit important ones

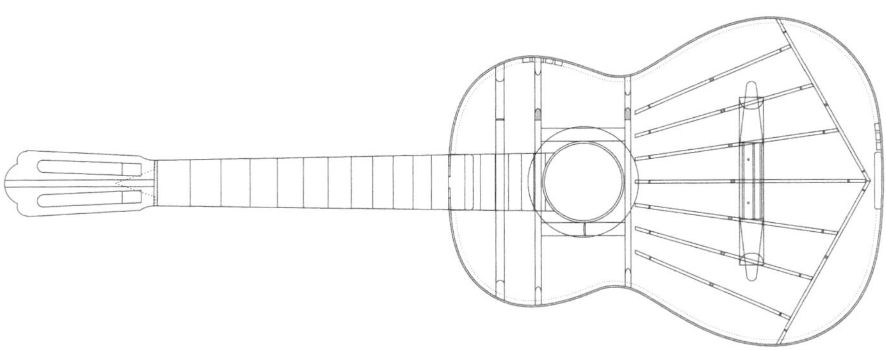

Fig. 3.1 Plan of 1937 Hauser guitar (image courtesy of Guild of American Luthiers, luth.org, plan by Richard Bruné, rebrune.com)

such as a much thicker top, distinguish it from the 1888 Torres appearing in the previous chapter. Figure 3.2 shows the front of the instrument.

Segovia was apparently pretty hard on his guitars and Bruné notes that the instrument showed signs of hard use. It has broken and missing peones (small triangular blocks that help fix the soundboard to the rim) and shows evidence of substandard repairs. It also was damaged by a falling microphone during a recording session and repaired by Hermann Hauser II, who sanded and refinished the top in the early 1960s. Segovia stopped playing this instrument shortly after the repair, saying that the first string had lost its superior tone.

Such is the importance of the instrument that many working luthiers now make reproductions of it. It should be noted that these instruments range from faithful copies to guitars that are probably more accurately described as being inspired by the original. Figure 3.3 shows a replica made by Kenny Hill in 2008 from Bruné's plan.

Table 3.1 shows basic dimensions of the instrument as reported by Bruné. The lutherie world has many examples of reported dimensions of historically important instruments differing slightly and so it is with this guitar. Various sources differ

Fig. 3.2 Segovia's 1937 Hauser (image courtesy of Richard Bruné, rebrune.com)

Fig. 3.3 Kenny Hill replica (image courtesy of Guitar Salon International, guitarsalon.com)

Table 3.1 Dimensions of Segovia's 1937 Hauser

Dimension	Metric (mm)	English (in.)
Nut width	50.8	2.0
Upper bout	279	10.98
Waist	235	9.25
Lower bout	365	14.37
Body length	485	19.09
Scale length (uncompensated)	650	25.60
Depth at heel	89	3.50
Depth at tail	98	3.86
Overall length	978	38.50
Soundhole diameter	86	3.39
Neck elevation	2	0.079
Headstock thickness	20	0.787
Neck thickness, first Fret[a]	21	0.827
Neck thickness, ninth Fret[a]	25	0.984
String spacing—outside centers	43	1.69

[a]dimensions measured from plan

slightly, a situation complicated by the fact that dimensions of this instrument appear frequently in plans, books, and web sites. The differences are usually small and could have many causes. One is that it's difficult to measure dimensions on an object with as few straight edges as a guitar. Another is that dimensions change slightly with changes in humidity. Because of the care taken in measurements and the detail of his plan, which was updated in 2016, I've stuck with Bruné's numbers.

Table 3.2 lists the materials from which the instrument was made. These are typical of what one would expect in a good classical guitar with the possible exception of the Indian rosewood bridge.

Table 3.2 Materials in 1937 Hauser guitar

Component	Species
Top	European Spruce
Back, sides, head veneer	Brazilian rosewood
Neck, head, heel, and all back braces	Honduras mahogany
Fingerboard	Ebony
Bridge	Indian rosewood
Glue	Hide glue

It's important to note that this instrument was made by hand and without the sophisticated fixturing one now routinely sees in guitar factories. It's not surprising then that the instrument is not perfectly symmetric. The top is surprisingly thick for a classical guitar. Contemporary instruments often had soundboards in the neighborhood of 2 mm (0.079 in.) thick, while the soundboard of this instrument ranges from 2.21 mm (0.087 in.) on the bass side of the lower bout to 3.24 mm (0.128 in.) on the treble side near the tail block. The top may have initially been thicker in some places before the repair by Hermann Hauser II.

The headstock is joined to the neck with a V joint, a common alternative to the scarf joint. Bruné notes that Hauser cut the V joint after the guitar was nearly complete. This may say something about Hauser's confidence as a builder. Engineers instinctively make complex objects in modules, bringing those modules together as late in the assembly process as possible. This way, an error in one of the components doesn't require discarding the whole thing. Hauser did the opposite by cutting the V joint after assembling the entire body with the neck as an integral part. Figure 3.4 shows a similar V joint, nicely executed in the headstock of a guitar by Greg Byers.

Some other features of this instrument distinguish it from other classical guitars of its time. They are subtle but may be important. The braces have rounded cross sections everywhere except the three center ones that Hauser left square underneath the bridge. The bridge itself is pinned to the top using 2 mm (0.079 in.) rosewood dowels and arched 1.5 mm (0.059 in.) across its width. The top was built flat and the final domed shape was imparted by the bridge as the top was glued to it. The wings of the bridge are thicker and, thus, stiffer than is typical.

A particularly interesting feature is a 1.5 mm thick cross grain lamination under the fretboard where it is glued to the soundboard. This appears to be an effort to reduce fretboard shrinkage, which can cause soundboard cracks next to the fretboard. This is not a feature widely adopted by later luthiers.

The neck of this instrument is not parallel to the plane of the soundboard. Rather, the end of the fretboard is raised approximately 2 mm higher than the plane of the top. This allows the strings to be closer to the soundboard, which reduces the torque on the bridge due to string tension, as shown in Fig. 3.5. Torque is force × distance, so reducing height reduces torque. String tension doesn't change unless tuning or scale length changes.

Fig. 3.4 A V joint in the headstock of a guitar (image courtesy of Greg Byers, byersguitars.com)

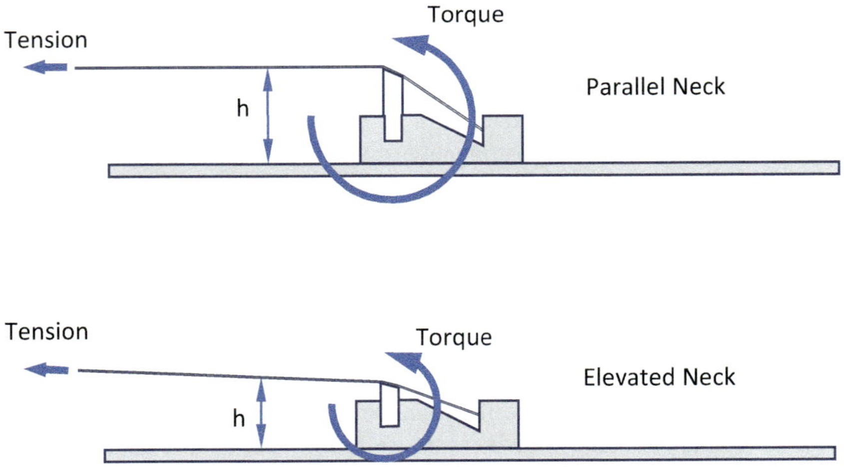

Fig. 3.5 Effect of reduced string height

Richard Bruné notes that torque is the #1 design feature in Spanish instruments and that a Spanish maker often starts with the height of the bridge and saddle, building the rest of the instrument to fit this.

The guitar is reported to be particularly resonant at G or 98 Hz. Bruné made several accurate copies of the original instrument and reports that they, too, have resonances near G. This could mean that its first body resonance is at G_2 (98 Hz), that its second body resonance is at G_3 (196 Hz) or perhaps both.

Other important features of this instrument that are worth studying, such as a slight relief in the fretboard that varies from treble to bass side. Rather than trying to describe them here, I encourage you to go to the full plan, which is quite detailed.

It's important that Segovia's widely reported opinion of this instrument is well corroborated. Julian Bream had an opportunity to play it and remarked that it was one of the finest guitars he had played. He made comments along these lines to both Jeff Elliott and Richard Bruné, who separately related it.

3.2 1933 Martin OM-28

The Martin OM (Orchestra Model) with a 14-fret neck was introduced before WWII. The 14-fret neck was developed in late 1929 after a suggestion to Frank Henry Martin by Perry Bechtel, a popular banjoist of the time. Previous designs had 12 fret necks and Bechtel reasoned that the additional two frets would make it a more versatile instrument. The result was one of the most successful acoustic guitar designs ever.

Figure 3.6 shows a 1927 Martin OOO-18 with the original OOO/OM body and a 12-fret neck. Note also the slotted headstock. Figure 3.7 shows a 1933 Martin OM-45 with a 14-fret neck. The 1927 instrument has what now looks like an extended upper bout, though this was the original Martin OOO/OM design. The upper bout on the 1933 instrument was shortened so that the neck met the body at the 14th fret.

The list of Martin Guitar models is dauntingly long. Their web site lists 18 body shapes for six string guitars with 12 or 14 fret necks and there are usually several versions for each body shape. The OM and OOO are very similar designs and model designations are sometimes written together as OM/OOO. OOO and OM models use the same body, and the major difference between the two is neck width and scale length. The OM has a longer scale length and a wider neck for fingerstyle playing.

Martin identifies most of its guitars with a body shape followed by a number designating its style or trim level. These trailing numbers ranged from 15 to 45, corresponding to trim levels. In this system, an OOO-18 was a lower priced OOO body guitar and an OM-45 was a top of the line OM instrument. The modern Martin line has expanded beyond this older system, so there are now other designations as well. The OOO-18 in Fig. 3.6 has a simple rosette, simple body binding, and an unbound neck, all features that reduced production costs. Conversely, the OM-45 in Fig. 3.7 is much more elaborate. It has mother of pearl or abalone in the binding and rosette. The neck is bound and the fret markers are much more elaborate than the simple dots in the OOO-18.

Fig. 3.6 A 1927 Martin OOO-18 with a 12-fret neck (image courtesy of C.F. Martin & Co. Archives, martinguitar.com)

There have been many different models in the OM/OOO line, but the OM-28 was one of the most popular. A modern version of it is still in production (Fig. 3.8) and it seems unlikely that the Martin line will ever be without an OM-28 model.

The OM and OOO had the right combination of tonal quality and versatility to draw wide acceptance. It's common to think of them as instruments for country and folk players, but it found its way into other genres. Figure 3.9 shows Big Bill Broonzy, an American blues singer, songwriter, and guitarist who was active from 1927 to 1958, with his Martin OOO-28.

Martin OM/OOO guitars made before WWII are particularly prized among musicians and collectors. An extreme example is the 1939 Martin OOO-42 that Eric Clapton played in his famous MTV Unplugged session. It was purchased by a collector in 2004 for $791,500 and certainly worth more now.

Fig. 3.7 A 1933 Martin OM-45 with a 14-fret neck (image courtesy of C.F. Martin & Co. Archives, martinguitar.com)

The OM is very popular among builders and numerous plans exist for it. A good one was drawn by Don MacRostie after careful measurements of a 1933 OM-28. Figure 3.10 shows the soundboard and bracing.

It says something about the popularity and the universal acceptance of this instrument that reproductions are available as kits from several suppliers, including Martin itself. Figure 3.11 shows a kit offered by Stewart MacDonald, based on the MacRostie plans.

Martin designs were seldom static, but constantly being refined. Thus, guitars with the same model numbers but produced in different years may have slightly

Fig. 3.8 A modern Martin OM-28 (image courtesy of Martin Guitars, martinguitar.com)

Fig. 3.9 Big Bill Broonzy playing his Martin OOO-28 (widely distributed, original source unknown)

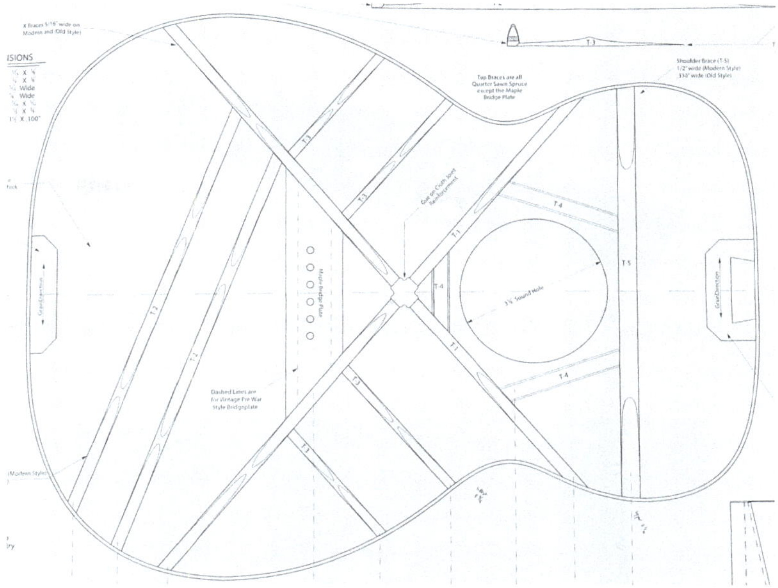

Fig. 3.10 Portion of OM-28 plan by Don MacRostie (courtesy of Stewart MacDonald, stewmac.com)

Fig. 3.11 A kit for a Martin OM reproduction (courtesy of Stewart MacDonald, stewmac.com)

Fig. 3.12 A reproduction Martin pyramid bridge (image courtesy of Stewart MacDonald, stewmac.com)

Fig. 3.13 A Martin Guitar made before 1867 with pyramid bridge (Wikimedia Commons, commons.wikimedia.org, image donated by the Metropolitan Museum of Art, metmuseum.org)

different dimensions or design features. Thus, there is no single OM design. One example is the bridge. Before about 1930, the wings had facets that led them to be called "pyramid bridges" (Fig. 3.12). Figure 3.13 shows a pre-1867 Martin Guitar with a pyramid bridge.

Fig. 3.14 A reproduction of a later Martin Bridge (image courtesy of Stewart MacDonald, stewmac.com)

Table 3.3 Dimensions of Martin OM-28

Dimension	English (in.)	Metric (mm)	Source
Nut width	1 3/4	44.5	Martin
Upper bout	11 13/32	289.7	Martin
Waist	9 3/16	233.4	MacRostie (measured from plan)
Lower bout	15 1/4	387.4	Martin
Body length	19 3/8	492.1	Martin
Scale length	25.4 25.34 24.465	645.2 643.6 646.8	Martin (uncompensated) MacRostie (treble end of saddle, front) MacRostie (center of saddle)
Depth (heel)	3 1/4	82.6	Martin/MacRostie
Depth (tail)	4 1/8	104.8	Martin/MacRostie
Overall length	39 1/2	1003.3	Martin
Soundhole diameter	3 7/8	98.4	Martin/MacRostie
String spacing at bridge	2 5/32 2 5/16 2 3/8	54.77 58.74 60.33	Martin (current) MacRostie (1933 Inst.) MacRostie (pre-1930 inst.)
Headstock thickness	5/8	15.9	MacRostie
Neck thickness at first fret	7/8	22.2	MacRostie (includes fretboard)
Neck thickness at eighth fret	31/32	24.6	MacRostie (includes fretboard)

In 1929, Martin introduced the bridge design that they still use on many of their instruments (Fig. 3.14), often called a "belly bridge."

Table 3.3 shows the dimensions of the 1933 Martin OM-28. This instrument was likely designed in fractional inches, so the English dimensions are reported that way. At this writing, the American guitar industry still works on decimal inches, while most of the rest of the world has long since gone to metric units. Note that some dimensions changed slightly as the design was refined. Dimensions here should be treated as nominal.

The different scale lengths in this table need some explanation. The 25.4 in scale listed on the Martin web site is the nominal dimension and the one used to locate the frets. Since the saddle has compensation (intonation) and is not perpendicular to the center line of the instrument, there is no single number that describes its distance from the nut. The face of the treble end of the saddle is 25.34 in from the nut and the center of the saddle is 24.47 in from the nut. Chapter 5 describes compensation in more detail.

Guitars sometimes have enough adjustments, compensation and even errors built into them that measuring the scale length is surprisingly difficult. The most well-known approach is to measure the distance from the inside face of the nut to the center of the 12th fret and double that length. This works if a short list of assumptions is valid, but they often aren't.

The problem is more difficult for older instruments. For reasons explained in a later chapter, fret locations of modern guitars depend on $\sqrt[12]{2}$ (the 12th root of 2, roughly 1.0595), as part of an expression called the ideal string equation. Those made before the 1920s often used a different, more approximate calculation. A further complication is that one can't always assume the frets are placed accurately. The widespread availability of precision measuring tools and CNC equipment is recent and older instruments can have slightly misplaced frets.

For a modern instrument without fret spacing modifications, a reliable method is to measure between the crowns (centers) of two widely spaced frets and calculate the resulting scale length. The popular method of measuring from inside face of the nut (the side facing the bridge) to the crown of 12th fret and then doubling that length works if

- The nut is placed without any shift to adjust pitch.
- The frets are placed accurately.
- Fret locations are calculated using $\sqrt[12]{2}$.

However, some builders shift the nut toward the saddle slightly to improve intonation. A method that works whether or not the nut has been shifted is to measure the distance from the 1st to 12th frets, divided by 0.44387.

$$L_{\text{scale}} = \frac{L_{12} - L_1}{0.44387}$$

This also assumes the frets are accurately placed using the ideal string equation and $\sqrt[12]{2}$. R. M. Mottola suggests that it's reasonable to assume that factory-made instruments from the 1960s onward have frets located this way. For older instruments or those made by hand, finding the design scale length can be an involved process (see Mottola's article in American Lutherie #136).

Some of the features of the Martin OM instrument have become nearly universal. In particular, the X-braced top is the default design for flattop, steel stringed acoustic guitars. The overall body shape and the 14-fret neck are equally familiar. That said, modern design has moved on from some of elements in this guitar.

Fig. 3.15 X brace with patch in a Guild dreadnought

The most popular change may be the bolt on neck. The dovetail joint on this and many other guitars of the time is strong, but problematic. It is geometrically complex enough to be a challenge for new builders, though not a problem given some experience and the right jigs. It also complicates the problem of resetting the neck, since the glue joint has to be steamed open.

Wood creeps under load and the neck alignment of older guitars can change enough that the neck needs to be reset. The glued dovetail joint needs to be steamed open, then the heel or the block in the body need to be modified to give the right angle. This can mean removing material or inserting shims. The joint then needs to be re-glued. A bolted neck can simply be removed and re-attached as needed.

Another, smaller change in some modern instruments concerns the center joint in the X brace. The braces need to be notched where they cross. The one notched at the top has greatly reduced stiffness. The stiffness of a beam (like a top brace) is proportional to the cube of its height. Reducing its height by half, then gives it 1/8 of its original stiffness. The cloth patch often glued over the joint is, structurally speaking, decorative. Figure 3.15 shows a cloth patch in the bracing of a Guild dreadnought guitar. It's now common to replace missing brace material with a small wood cap strip that restores the missing stiffness.

3.3 1948 D'Angelico New Yorker

Orville Gibson is usually credited with the development of the archtop guitar, though by way of the archtop mandolin. In 1898, he was awarded US Patent 598,245 for an archtop mandolin. Figure 3.16 shows the first page with a drawing of a mandolin with an arched top and back. Up to that point, mandolins commonly had bowl backs.

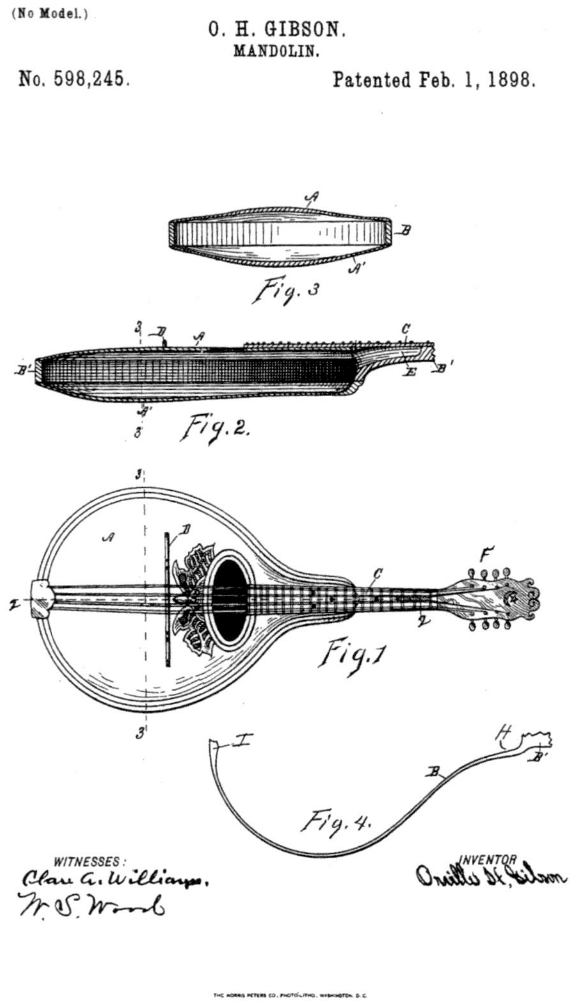

Fig. 3.16 Orville Gibson's 1898 Patent for an archtop mandolin

Patents usually make for some pretty dry reading, but this one has a surprising rhetorical flourish

> …I have attained that degree of success with continued experiments and manufacture that every portion of the woody structure seems to be alive with emphatic sound at every touch of the instrument—a character and quality of sound entirely new to this class of musical instruments, and which cannot be imparted to others by description in words.

His enthusiasm for the arched top apparently couldn't be contained to the mandolin, as he introduced the L-1 archtop guitar in 1902. Somewhat confusingly, Gibson reused the designation L-1 in 1926, for a flattop model famously played by Robert Johnson.

Gibson refined archtop designs and introduced the L-5 in 1922 under the direction of Lloyd Loar. Versions of it have been in production ever since. The L-5 was originally introduced as an acoustic instrument, but was later fitted with an electromagnetic pickup.

John D'Angelico opened his workshop in 1932 and started by making instruments similar to the L-5. His early work is important enough that the Metropolitan Museum in New York has a 1932 D'Angelico in its collection. Figure 3.17 shows a 1937 New Yorker. Its relationship to the L-5 is clear, though D'Angelico's own aesthetic had emerged. It's clear that this is a very large instrument. Two of the most widely known D'Angelico designs were the Excel (17 in body) and the New Yorker (18 in body).

Fig. 3.17 A 1937 D'Angelico New Yorker (image courtesy of Rudy's Music, rudysmusic.com)

Fig. 3.18 1960 D'Angelico New Yorker (image courtesy of Rudy's Music, rudysmusic.com)

His New Yorker model became one of his most well-known and eventually incorporated the art deco elements that came to define his work. Figure 3.18 shows a 1960 D'Angelico New Yorker fitted with an aftermarket electromagnetic pickup. The pickup is at the end of the neck, but a rod along the bass side of the strings leads to a fitting that supports the controls. D'Angelico made about 300 New Yorkers in his career, a small enough number that they are prized by collectors.

D'Angelico died in 1964 and his apprentice, Jimmy D'Aquisto, went on to establish himself as a maker of fine archtop guitars. D'Aquisto bought the D'Angelico business after the founder's early death, but lost rights to the name as the result of an unfortunate business deal. D'Aquisto went on to make guitars under his name

Fig. 3.19 A modern D'Angelico Excel EXL-1 (image courtesy of D'Angelico Guitars, dangelico-guitars.com)

until his own early death in 1995 (see Acquired of the Angels by Paul William Schmidt). The D'Angelico name was later purchased by investors and a new company, D'Angelico Guitars was launched in 2011. They offer a guitar based on the original Excel (Fig. 3.19), but not one based on the New Yorker. With the 18 in. lower bout, it is a very large instrument and may not be practical for enough players to justify production.

Original D'Angelico archtops are ornate instruments and their art deco aesthetic is part of what makes them iconic. Figure 3.20 shows the signature D'Angelico "Stair Step" style tailpiece on a 1949 Excel.

Fig. 3.20 An original D'Angelico "Stair Step" tailpiece (image courtesy of D'Angelico Guitars, dangelicoguitars.com)

John D'Angelico seems to have continually modified his designs, so exploring the details of one specific instrument gives insight into his thinking at the time he made it. In 1988, luthier Steve Andersen received a 1948 New Yorker for replacement of the binding around the top and back. While working on the instrument, he drew a plan of it (Fig. 3.21, GAL Instrument Plan #24 – 1948 D'Angelico New Yorker). In preparing the plan, he contacted Jimmy D'Aquisto to verify some of the details.

This is a very large instrument with the familiar X-braced top, f-holes, and a floating pickguard. The top is made from what Anderson describes as unmatched Sitka spruce plates with wide, wavy grain. The back and sides are made of maple with a light bird's eye pattern. It is finished with lacquer in a sunburst that goes from amber at the center through brown and to black at the edges (Fig. 3.22).

Table 3.4 lists the dimensions of the instrument. Note that these dimensions were measured from Andersen's plan and should be taken as nominal.

This is an ornate instrument, with elaborate binding on the body, neck, and headstock. Figure 3.23 shows the headstock of a 1941 New Yorker with the trademark D'Angelico inlays and binding.

Fig. 3.21 Plan of a 1948 D'Angelico New Yorker (image by Steven Andersen, used courtesy of the Guild of American Luthiers, luth.org)

Fig. 3.22 A 1948 D'Angelico New Yorker (image by Steven Andersen, courtesy of the Guild of American Luthiers, luth.org) (1988)

Table 3.4 Dimensions of 1948 D'Angelico New Yorker

Dimension	English (in.)	Metric (mm)
Scale length	25 5/16 (at Treble E string)	642.9
Overall length	44	1118
Body length	21 7/16	544.5
Upper bout width	13 7/8	352.4
Waist	11 9/16	293.7
Lower bout width	18 1/8	460.4
Headstock angle	15 ½°	N/A
Neck angle	3 ½°	N/A

Fig. 3.23 Headstock of a 1941 D'Angelico New Yorker (image courtesy of Rudy's Music, rudysmusic.com)

Note that D'Angelico used at least two different headstock designs on his instruments. Figure 3.24 shows the headstock on a 1960 New Yorker. Modern guitars by the new D'Angelico company use a headstock design similar to this one.

The art deco inlay on the headstock matches the stepped tuners knobs. This style of knob is still popular enough that Grover offers tuners in the same style as the original (Model 150 Imperial, Fig. 3.25).

Fig. 3.24 Headstock of a 1960 D'Angelico New Yorker (image courtesy of Rudy's Music, rudysmusic.com)

Fig. 3.25 Grover Model 150G Imperial tuners (image courtesy of Sweetwater Sound, sweetwater.com)

Fig. 3.26 A 1960 Gibson ES-335TD semi-hollow body (Wikimedia Commons, commons.wikimedia.org, image uploaded by user Lubbad85)

The 1948 New Yorker is a true hollow body acoustic guitar. This is an important distinction that separates it from semi-hollow body guitars that have additional interior structure joining the top and back along the center of the body. This internal structure limits feedback and provides a place to mount a bridge and tailstop. Figure 3.26 shows a Gibson ES-335 semi-hollow body guitar. It has a Tune-o-Matic style bridge with tailstop and both have studs mounted directly into the core. True hollow body archtops don't have the center core, so they use a floating bridge and tailpiece (Fig. 3.26).

Bibliography

Books

Johnston, Richard and Boak, Dick, Martin Guitars: A History, Hal Leonard 2008, ISBN 978-0-634-03785-6

Johnston, Richard and Boak, Dick, Martin Guitars: A Technical Reference, Hal Leonard 2009, ISBN 978-1-4234-3982-0

Osborne, Nigel (creative director), The Classical Guitar Book, Backbeat Books, 2002, ISBN 978-0879307257

Schmidt, Paul William, (n.d.) Acquired of Angels: The Lives and Works of Master Guitar Makers John D'Angelico and James L. D'Acquisto, 3rd ed., DHR Publishing and Articulate Passion LLC, ISBN 978-0692165515.

Schmidt, Arian and Sheets, Paul, The Master's Bench: The Guitar Making Workshop of D'Angelico, D'Acquisto and Gudelsky, National Music Museum, 2016, ISBN 978-0984826940

Articles

Anderson, Steven, 1948 D'Angelico New Yorker, American Lutherie, #16, 1988, Also in Big Red Book of American Lutherie 2, ISBN 0-9626447-3-0.

Bruné, Richard, Segovia's 1937 Hauser: Top and Back Thicknesses, American Lutherie #38, p. 24, 1994, also in the Big Red Book of American Lutherie 4, ISBN 978-0-9626447-6-4.

Bruné, Richard, Andre Segovia's 1937 Hauser Built in 1937, Vintage Guitar, October 2003.

Bruné, Richard, 1937 Hermann Hauser Sr. Classic Guitar, Ex Segovia, American Lutherie #126, p. 34, 2016a

Bruné, Richard, GAL Instrument Plan #33 - 1937 Hermann Hauser Sr. Classic Guitar, Ex Segovia, American Lutherie #126, p. 36, 2016b

Elliott, J. and Burton, C., Reflections on Segovia's Guitars at the Metropolitan, American Lutherie #21, p. 32, 1990, also in the Big Red Book of American Lutherie, Vol. 2, ISBN 0-9626447-3-0.

Krick, George C., Fretted Instrument Department – The Classic Guitar, The Etude Magazine, Vol. 55, No. 12, December 1937

Segovia, A., In Memoriam – Hermann Hauser, Guitar Review, No. 16, 1954

Mottola, R.M., (n.d.-a) Measuring Scale Length of Fretted Instruments, American Lutherie #136, p. 48

Mottola, R.M., (n.d.-b) Was the Rule of 18 Good Enough?, American Lutherie #130, P 52

Chapter 4
Basics of Guitar Design

Guitar makers are driven by many different things, but we enjoy and often need the satisfaction of bringing an object into being. Working from existing designs makes us craftsmen. The artisans among us are those who go a step further to start with a vision rather than a design, moving from the honorable work of making a fine object to the aspiration of creating something uniquely our own.

Andy Powers on Guitar Design

It seems to me that to make something well is to achieve a balance of complementary imperfections. To refine it is a study in diminishing returns. A musical instrument requires both a design, and ability to give form to the design. Driving this endeavor is a blend of understanding and inspiration. One must pursue understanding of the mechanisms at play and relationships between individual pieces, their look and function. As well, we must respect and seek ever higher levels of craftmanship to execute a design. In the same way that a musician is creating and responding to sounds, a great craftsperson knows their materials and is both guide and follower as those materials are shaped into an instrument. Despite all the effort which goes into creating a guitar we must remember the instrument, a compelling piece of inspired art itself, must remain in service to the musician. After all, a guitar doesn't make a sound until it is being played, and ultimately, the measure of a great instrument is one that a musician enjoys playing.

There are many reasons one might want to design a guitar. The motivation may be musical: to make a guitar to fit the needs of a specific player or to give voice to a style of music. It may be aesthetic: to create a guitar as an object of beauty. It may be exploratory: to just try some idea to see if it works. There is no wrong answer. We

R. M. French, *Acoustic Guitar Design*,
https://doi.org/10.1007/978-3-030-89381-1_4

are free to design instruments just because we want to and it's here that the most creative freedom lies.

Whatever else it might be, a guitar must first be a musical instrument that pleases the player. Physics and the requirements of the musical scale present the guitar designer with inviolable boundaries that other artists don't face. You can't cheat either; Nature will always call your bluff.

Other practical limits constrain the designer and these can be quite demanding. Cost is an obvious example that is seldom far from most designers' minds. Indeed, the guitar world is full of stories about people who, faced with difficult circumstances, were resourceful enough to make and to master instruments made from whatever materials came to hand. Figure 4.1 shows Moses Williams (1919–1988) playing a large diddley bow using a bottle as a slide. The instrument is little more than a plank (it looks like it could have originally been a door or a shelf) with a wire stretched between nails or screws and tensioned with an old can. It would be quite a reach to call this an acoustic guitar, but it had the same job and it helped Williams to find his unique musical voice. In that sense, it was a successful instrument.

Fig. 4.1 Moses Williams playing a large diddley bow in 1978 (Florida Memory—State Library and Archives of Florida, image is in the public domain)

Fig. 4.2 Pat Metheny playing the Pikasso guitar by Linda Manzer (Wikimedia Commons, uploaded by PaulCHerbert, Creative Commons License)

Sometimes the imagination of the designer is so ambitious that it operates near the limits of physics or human physiology. Figure 4.2 shows the famous jazz guitarist Pat Metheny playing his Pikasso guitar, made by Linda Manzer. Manzer is among the top luthiers now working and has established herself as an artisan of great creative range. Whatever else the Pikasso is, it is first a fine guitar. Though, one wonders whether the first step in learning to play it is to be Pat Metheny. Somewhere between these two creative extremes is the space where most of us work. Fortunately, there is plenty of room.

It helps to have a basic process to follow when designing a guitar. Figure 4.3 shows one process modeled on those commonly used in the engineering world. The four steps in this approach are: concept, basic design, detail design, and fabrication. Each box lists some of the choices the designer needs to make for that step. It's important that this process is recursive. It's quite rare for the first version of a new design to need no revisions. Many builders intuitively work through a process like this one in an informal way and designers often loop through this process more than once before being satisfied with the result. In particular, it's common to find problems when making a prototype that require modifications to the design. Often the changes are only in details, but they can still require another pass through the design process.

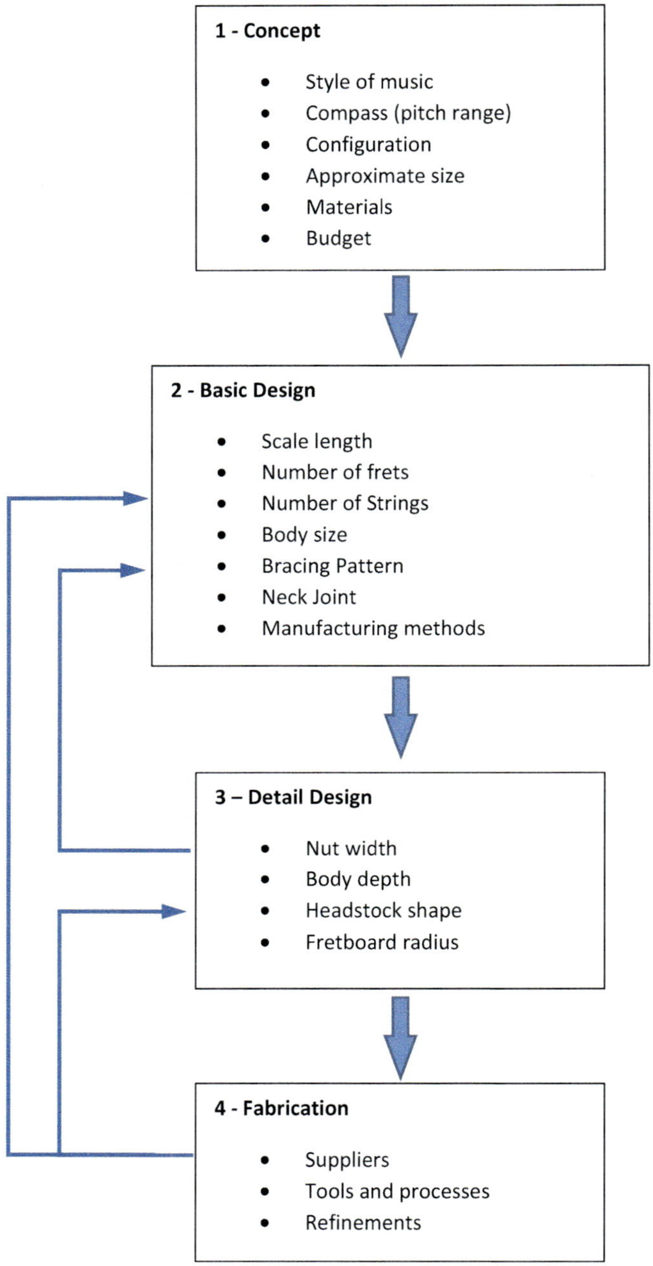

Fig. 4.3 Sample process for designing an acoustic guitar

4.1 Design Concept

The concept stage is where the designer has the most freedom. This is the time to think crazy thoughts and to consider unconventional solutions if you like, since practical considerations will start to take over after this.

The most basic design decisions to make are usually what kind of music the instrument is intended for and who the intended player is. These are often two facets of the same question. For example, a classical player is unlikely to want an archtop guitar, though it is certainly possible to make an archtop, nylon string guitar with a wide neck for classical music. Figure 4.4 shows such an instrument, made as an experiment.

Fig. 4.4 An experimental nylon stringed archtop guitar

Kenny Hill on Guitar Design

I've played a lot of guitar steadily since I was a kid, and my design evolution has roughly tracked my own needs, desires and progress as a musician. There are certainly decisions that get shaped by economics and shifting trends, but mostly I look to satisfying myself first, and hoping that the musical buying public will have complimentary tastes to my own. It took me a long time to give myself permission to consider that my own thoughts might be as legitimate as those of my predecessors and heroes, and just possibly might be even better.

I eventually figured out that there are no defining secrets and there's no single way to make a good guitar. And every instrument doesn't have to be your masterpiece. Different players need different guitars. Bracing style, body shape, wood, finish and construction process combine in many different ways and result in some kind of balance between all of the elements. Most of the historic icons of guitar making were really just gifted tradesmen trying to make a living. And most of the historic designs that are now enshrined were iconoclastic breaks with the tradition of their day, and while there is much to be learned from the past, musicians' needs and guitar fashion are always changing, and there are no final right answers.

Physics is sometimes less restrictive than the expectations of players. For example, classical players, particularly those with formal training, usually prefer guitars that follow a specific, familiar pattern. Similarly, jazz players often expect an archtop guitar that reminds them of a D'Angelico, a D'Aquisto, or a Benedetto. The finest guitar in the world is of no use unless people are willing to pick it up and play it.

Designers must keep in mind the intended pitch range, or compass, of the instrument. Most guitars are designed with standard EADGBE tuning mind, though players sometimes use alternate tunings. Standard tuning means that the lowest open string frequency, the bass E, is 82.4 Hz and highest open string frequency, the treble E, is 329.6 Hz. Baritone guitars are tuned lower, often five half-steps, called a perfect fourth. Small-bodied guitars are often tuned higher. For example, the small Terz guitar is tuned a minor third, three half steps, higher than standard tuning.

Another important first question about the guitar is how big it should be. Figure 4.5 shows the body of a 1941 D'Angelico New Yorker. It's beautiful and there are many good reasons to want one. However, with an 18 in. (457 mm) wide lower bout, it's a beast of a guitar. Smaller players would likely struggle with it and it is proportionally more difficult to travel with. It seems obvious that the guitar must be sized for the intended player as part of its intended job. A guitar for smaller players needs to be scaled down and there are a number of good ¾ sized instruments now available.

If the guitar needs to be highly portable, it helps to make it smaller. Being able to fit into the overhead storage bin on an airplane is sometimes a critical design

Fig. 4.5 A 1941 D'Angelico New Yorker (image courtesy of Rudy's Music, rudysmusic.com)

requirement and has driven a whole class of travel sized guitars. However, basic physical laws dictate that, up to a point, larger instruments tend to sound better.

The two most common complaints about travel guitars are that they don't balance well and they have little low frequency content—the sound is often described as trebly or thin. The design process is always a matter of balancing conflicting requirements and most travel guitars are built for easy portability at the expense of tonal quality.

Figure 4.6 shows an experimental travel guitar. The two top design requirements were that it had to fit in the gig bag for a Martin Backpacker and had to balance well. The Backpacker is quite successful; it is very small and portable, so it is easy to carry around. It's also durable, so it can handle the bumps and dings that come with being stuffed into backpacks and airplane overhead bins. However, it has poor balance and players sometimes complain about its tone.

Fig. 4.6 An experimental travel guitar

This experimental instrument has a titanium truss rod and Gotoh Stealth tuners to keep the neck light. These features, combined with a heavy tail block, let it balance naturally on a strap just like a full-sized instrument. It also fits into a Backpacker gig bag. The sound quality is fine for a travel guitar, but not comparable to a conventional acoustic guitar. Later versions may have better sound quality.

The most general rule is that larger guitars have better low frequency response, though skilled designers can sometimes coax good low frequency sound from smaller guitars. The Taylor GS Mini (Fig. 4.7) is one of the most successful small guitars, partly because of its good tone. It has a scale length of 23.5 in. (597 mm) and is classed as a ¾ size instrument. The ¾ designation is notional does not mean that it is 75% as large as a full-size guitar. For example, Taylor uses a 25.5 in. (747.7 mm) scale length on most of its full-size instruments and ¾ of that is 19.125 in. (485.8 mm).

Fig. 4.7 Taylor GS Mini (image courtesy of Taylor Guitars, taylorguitars.com)

Another of the decisions to make early on is which materials to use. A large majority of guitars are still made from wood and few individual builders use other materials, except for reinforcement. Wood is inexpensive, familiar, and easy to work. More importantly, guitars have always been made of wood, so players instinctively expect their instruments to be made with it. Composites are slowly working their way into the market, but they require different tools and processes. For now, composite instruments are mostly the domain of manufacturers rather than individual luthiers.

Budget is a consideration at the concept stage, particularly if the instrument is for a paying client. For an inexpensive guitar made by an individual luthier, materials may be as much as 1/3 of the sale price. For expensive instruments made by the best luthiers, material costs may be a smaller proportion. Few builders have the luxury of an unlimited budget, so designers are wise to spend money where it will do the most good.

For example, a cheap set of tuners is, at this writing, \$12–\$15, but they are usually of poor quality. Apart from causing tuning problems, they can also turn away potential buyers. Conversely, there is little practical difference between a good set of steel string tuners from a reputable manufacturer (Grover, Gotoh, Schaller and a few others) and a hand-made custom set. A good set of production tuners for steel string guitars is now around \$50–\$100, while boutique sets can easily cost several times more. There are few mechanical reasons to spend lavishly on tuners, but customer expectations may drive the decision.

Exotic wood with captivating, figured grain can be breathtakingly expensive, but has little effect on tone when used for the backs, sides, and decorative elements like headstock plates. However, it can attract buyers who want a truly unique instrument. At this writing, a nice, but unremarkable back and side set for a full-sized acoustic guitar may cost \$100–\$150. At the far opposite end of the scale is wood whose origin (and cost) borders on the mythical.

In 1965, a mahogany tree was felled in the Chiquibul rainforest whose grain was highly figured. A team led by Robert Novak eventually extracted it from the ravine

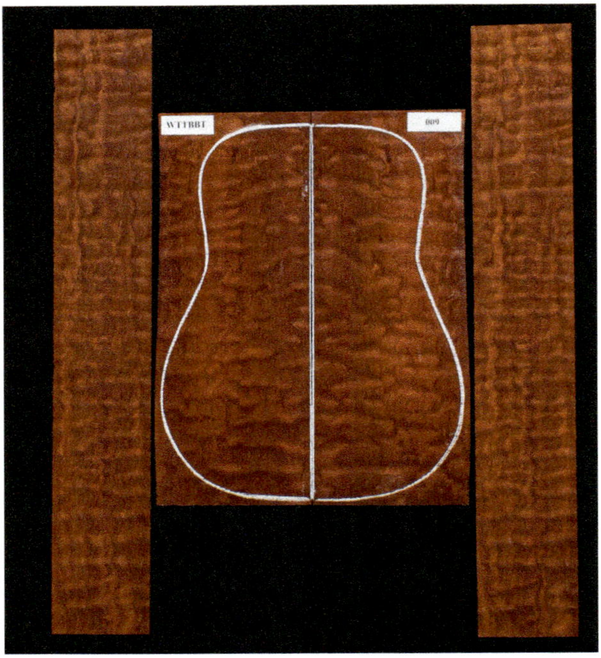

Fig. 4.8 Back and sides from The Tree (image courtesy of Stewart MacDonald, stewmac.com)

into which it fell and had it processed into wood for guitars. Now widely known as The Tree, back and side sets are still available, though at prices that will scare off all but the most committed builders. Prices in excess of $3000 for a back and side set from this log are common. Figure 4.8 shows a beautiful set from The Tree.

4.2 Basic Design Parameters

The design of acoustic guitars is dominated by a few basic requirements:

- Scale Length
- String Tension
- Neck Joint
- Neck Structure
- Neck Shape
- Body Shape
- Materials

There are others, but these are the ones that we must address first as they will drive much of the rest of the design. As a guitar designer, you have freedom to express yourself, but only within the design limits that will give you a successful guitar. Let's start by establishing some of those limits.

4.2.1 Scale Length

The number that most defines the overall size of the guitar is scale length. In theory, scale length could be just about anything. In practice, it is limited by the size of the player and the dynamics of stretched strings. For a guitar, the shortest commonly used scale length is about 22 in. (558.8 mm), though there is a small group of six string ukulele sized instruments (guitarleles) with 17 in. (431.8 mm) scale lengths. The largest common scale length for a standard guitar is about 26 in. (660.4 mm), with longer ones typically reserved for baritone guitars. Table 4.1 shows the range of commonly used scale lengths.

The upper end of the range is defined by ergonomics since people simply can't play a guitar that is too big. The lower end of the scale is set by both ergonomics and the physics of vibrating strings. The ergonomic problem is obvious since, when frets are too close together, chords become difficult to play. A more subtle problem has to do with how stretched strings vibrate.

Vibrating strings make more than one frequency at a time. The lowest one is called the fundamental. The higher ones, often called harmonics, should be multiples of the fundamental. For example, the A string in standard tuning should have a fundamental frequency of 110 Hz. The higher frequencies should be 220 Hz, 330 Hz, and so on. For longer strings, this is close to correct. However, the shorter the string, the more of a problem this is.

The formal name for this problem is inharmonicity. Short strings can sound bad, even when the guitar is in tune, because the higher frequencies depart too far from the harmonic ideal. Increasing the string tension reduces the problem, and it helps that small guitars are often tuned to higher pitch. Thinner strings also reduce the problem since inharmonicity arises from bending stiffness of the strings. The mathematical model of an ideal string ignores bending stiffness and produces an ideal harmonic series.

Table 4.1 Common guitar scale lengths

Type	Length (in.)	Length (mm)
Guitarlele (tenor ukulele)	17	431.8
¼ size guitar	22	558.8
Child's guitar, travel guitar, Baby Taylor	22.75	577.9
¾ size guitar, Taylor GS Mini	23.5	596.9
Gibson J-45	24.75	628.7
Martin 00	24.9	632.5
Full size, Martin OM, Taylor GA	25.5	647.7
Long scale classical guitar	25.98	660
Taylor baritone	27	685.8

4.2.2 String Tension

Tension is a function of tuning, materials, and scale length, and it's not something the designer chooses directly. Rather, it's the result of other design decisions. Still, string tension is the primary load applied to the guitar and designers need least a rough idea of what the forces are. Let's assume we're using a 25.5 in. (647.7 mm) scale length and standard EADGBE tuning. There are many alternate tunings, but few of them increase the tension very much. Probably the most common alternate tuning is drop D, which only slightly decreases the total tension. Increasing pitch of all the strings could significantly increase tension.

Strings are usually sized so that the tension in a set of strings is roughly uniform from string to string. Table 4.2 lists the tension of representative sets of strings. Engineers and scientists among us will reflexively note that kg is a unit of mass, not of force. They are right, but it doesn't matter; the world (incorrectly but obdurately) uses kg as a unit of weight and force.

4.2.3 Neck Joint

All guitars have a body and neck, so one of the most important structural features is the joint connecting them. While there are many variations, there are basically three types:

- Dovetail
- Bolted
- Spanish Heel

Dovetails are simple, require no additional hardware, and have certainly proven to hold up over time. However, they can require careful fitting when not made using production tooling and they make neck resets difficult. Wood permanently deforms over time (creeps) so it's not uncommon for the necks of older guitars to have raised

Table 4.2 Representative string tensions

Description		E	A	D	G	B	E	Total
D'Addario EJ45 Pro Arte Normal Tension Classical	lb	14.19	15.89	15.62	11.88	12.04	16.23	85.85
	kg	6.44	7.21	7.08	5.39	5.46	7.36	38.93
D'Addario EJ40 Silk and Steel	lb	18.48	22.4	21.16	27.5	17.85	19.63	127.02
	kg	8.38	10.16	9.60	12.47	8.09	8.90	57.62
D'Addario EJ16 Phosphor Bronze Light	lb	24.95	28.93	29.93	30.06	23.31	23.36	160.51
	kg	11.32	13.12	13.58	13.63	10.57	10.60	72.81
D'Addario EJ17 Phosphor Bronze Medium	lb	27.75	32.77	35.87	35.15	26.31	27.42	185.27
	kg	12.59	14.86	16.27	15.94	11.93	12.44	84.04
D'Addario EJ18 Phosphor Bronze Heavy	lb	31.02	38.50	43.94	38.06	29.50	31.80	212.82
	kg	14.07	17.46	19.93	17.26	13.38	14.42	96.53

slightly. This increases the string height over the fretboard, sometimes by an unacceptable amount, so the neck needs to be removed and re-glued at the correct angle. This requires steaming the glued joint open. Changing the alignment often requires shaving material or inserting shims. Resetting a neck with a dovetail joint is a famously fussy job and a constant source of work for guitar repair people.

An alternative that is now very common in both production and in individually made instruments is a bolted neck joint. Most of the time, the neck is attached with bolts which mate to nuts or threaded inserts. Resetting a bolted neck joint is a simple matter of removing the neck, inserting shims or removing material to give the correct alignment and bolting the neck back on.

As mentioned in Chap. 2, Taylor Guitars developed a design in which the neck is aligned in the factory using spacers that are selected during final assembly to give the correct neck angle (Fig. 4.9). The neck cannot be assembled without these spacers and final assembly stations are provided with a range of them from which to choose. A neck reset is just a matter of replacing these spacers.

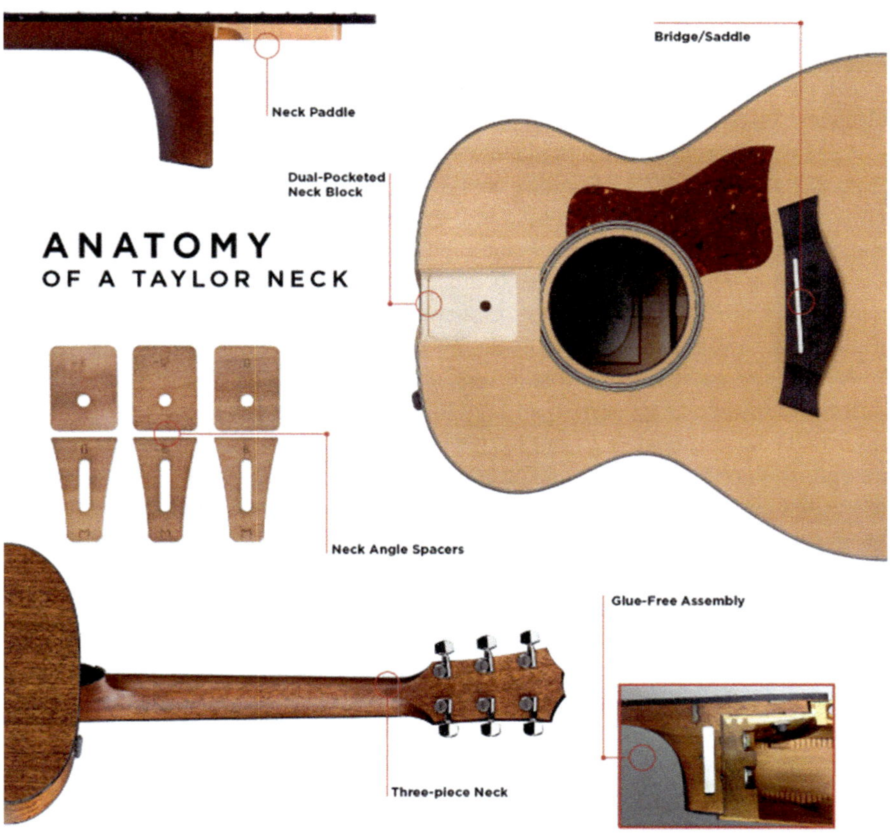

Fig. 4.9 The Taylor neck joint (image courtesy of Taylor Guitars, taylorguitars.com)

Fig. 4.10 A simple, robust bolted neck joint (image courtesy of R.M. Mottola, liutaiomottola.com)

Bolted neck joints don't need to be elaborate in order to be robust and accurate. Figure 4.10 shows a simple, strong bolted neck joint by R.M. Mottola. Clearly visible are two cap screws with washers and the nut for the truss rod.

An even simpler approach is an extension of the bolted neck joint used in Fender electric guitars. Popular terminology notwithstanding, these are not strictly bolted joints. They are actually held in place with sheet metal screws threaded into the heel from the bottom—1¾ in. #8 screws are typical for solid body electric guitars. While not very sophisticated, there is no doubt that this method works, having been used since the 1950s on almost all Fender guitars and others made on that pattern. Compressive loads from the screw heads are typically distributed by a rectangular neck plate or individual ferrules. A refinement is to install threaded inserts into the neck so that it is attached with bolts.

Millions of electric guitars have used this simple, robust design. If the screws pull out of the neck, it is easy to drill out the holes, glue in dowels to plug them, and re-drill. There is probably a limit to how many times this can be done. In practice, though, this joint lasts the working life of the instrument. Figure 4.11 shows the neck joint on a 1966 Telecaster. The guitar was 43 years old when this picture was taken and the joint appears to be holding up well.

Some acoustic guitars follow this approach, though with modifications to accommodate the larger body depth. If the designer is content with longer bolts, they can run from the back into the neck, just like on Telecasters and Stratocasters. A more direct approach is to put the screws through the fretboard and into the neck block. Taylor uses this design for their Baby and Big Baby guitars. While not, perhaps, elegant, it is simple, light, robust, and inexpensive.

The necks are held in place with a pair of screws, placed side by side between the 15th and 16th frets. Since the screws are black, they are nearly invisible against the black ebony soundboard. These instruments don't have a heel for the simple reason that they don't need one. Figure 4.12 shows a mahogany Baby Taylor BT2 with a

Fig. 4.11 The neck screws and plate on a 1966 Telecaster (Wikimedia Commons, commones. wikimedia.org, image by user ArtBrom)

Fig. 4.12 A Taylor BT2 Baby Taylor acoustic guitar (image courtesy of Taylor Guitars, taylorguitars.com)

light ebony fretboard (not all ebony is black) that makes the black neck screws easier to see. This simple, light neck joint has proven to be durable, having been used on thousands of Baby Taylors, since their introduction in 1996, and on Big Baby Taylors, since their introduction in 2000. If there was something wrong with this neck joint, we'd know by now. It's worth noting that the great master violin makers originally used nails to help secure the necks. By comparison, screws are an improvement.

4.2.4 Neck Structure

The neck is essentially a cantilevered beam that has to be strong enough to bear string loads, form a good dynamic boundary condition for the strings, and provide a comfortable place for the player's hand. The mechanical joint with the body holds the neck firmly to the guitar, which is what makes the neck a cantilever. In order, let's consider strength, dynamics, and shape.

The neck of a steel string guitar, strung with medium strings, must bear a load about equal to the weight of a grown man. Necks of classical guitars must bear about half of that. The good news is that almost any practical wood is strong enough not to break under that load. For example, the compressive yield strength of mahogany, a common neck wood, is in the neighborhood of 6500 psi or 44.8 MPa. Thus, in ideal conditions, a square stick of mahogany about 5 mm on a side (about 0.20 in.) would be more than strong enough. No set of guitar strings will break a guitar neck in compression unless there is some other structural problem.

Bending is more of a problem. While strings never break a neck in compression, it is quite possible for them to deform the neck enough to raise the strings unacceptably far above the fretboard. This is called up bow. Figure 4.13 shows the effect of too much neck deflection.

Since a set of steel strings can have about twice the tension of nylon ones, steel stringed guitars generally include neck reinforcement in the form of a truss rod.

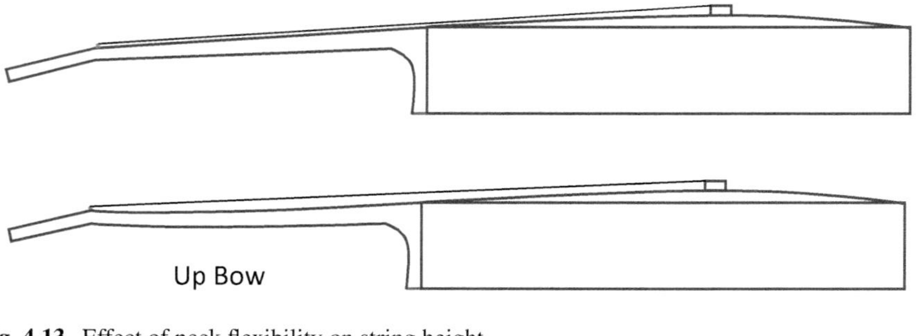

Fig. 4.13 Effect of neck flexibility on string height

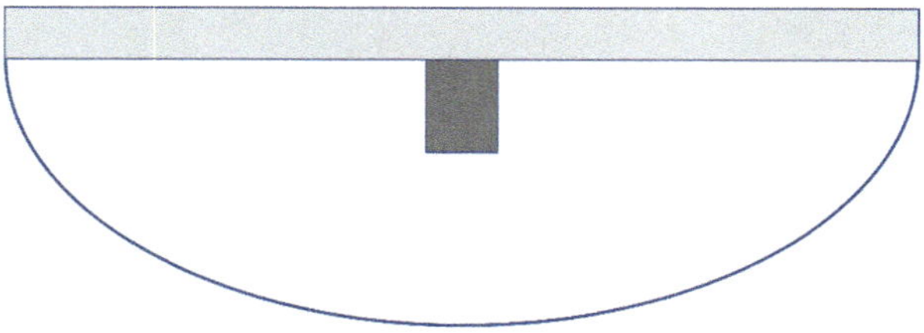

Fig. 4.14 Fixed truss rod

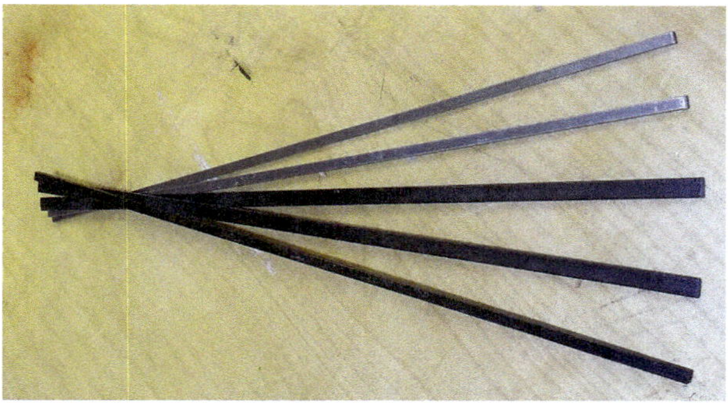

Fig. 4.15 Carbon fiber and titanium truss rods

Classical guitars, with their lower string tension and wider necks, sometimes have truss rods, but they are far from universal.

Truss rods come in two flavors, fixed and adjustable. A fixed truss rod is just what it sounds like—an inset bar that increases bending stiffness, but can't be adjusted (Fig. 4.14) to alter the curvature of the neck.

Fixed truss rods can be made out of any material stiff enough to substantially reduce deformation and light enough not to cause balance problems. Some traditional classical guitar makers use a piece of wood of a species stiffer that used for the rest of the neck. The early builders sometimes used sticks of ebony to stiffen necks and this is still occasionally used in classical guitars. For classical guitars fitted with fixed truss rods, it's now common to use carbon fiber (graphite). At least one lutherie supplier now offers titanium bars as well. Figure 4.15 shows a selection of carbon fiber and titanium truss rods.

Most modern steel string acoustic guitars include adjustable truss rods to allow for more precise setup, though they are more rare on classical guitars. Martin used fixed truss rods, starting with ebony bars around 1920 (Fig. 4.16). They replaced ebony with T-section steel bars in 1934, with ebony bars coming back during WWII due to a steel shortage. The T-section bar was replaced by square steel tube in 1967. Martin introduced an adjustable truss rod in 1985.

Necks can change shapes due to changes in humidity or changing string tension. An adjustable truss rod lets the player correct the curvature of the neck as needed.

Fig. 4.16 A Martin guitar neck with fixed ebony truss rod (courtesy of C.F. Martin & Co. Archives, original image by Ian Nansen)

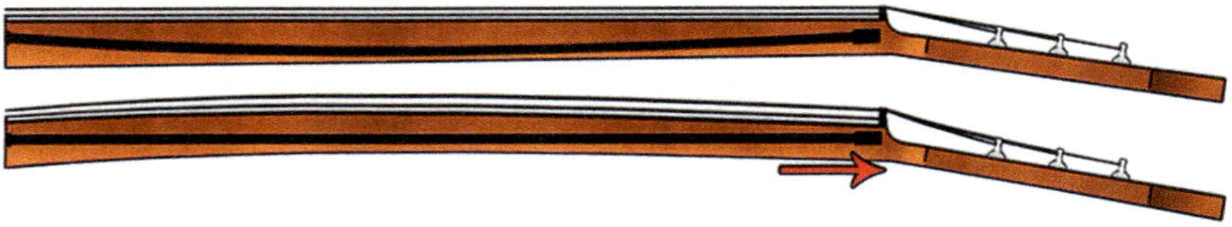

Fig. 4.17 A single acting truss rod (Wikimedia Commons, commons.wikimedia.org, created by user Bamnehagen)

Additionally, players typically prefer a slight up bow in the neck, called relief. A discussion of the relief appears later.

There are two types of adjustable truss rods, single acting and double acting. As the name suggests, a single acting truss rod can only induce a curve in one direction. A single acting rod, confusingly also called both a tension rod and a compression rod, is usually set into the neck with a slight curve, as shown in Fig. 4.17.

One end is fixed to the neck, usually with a threaded block or barrel nut, and the other is threaded so that a nut can put the neck in compression and the rod in tension (Fig. 4.18). Tension in the curved rod causes it to straighten, reducing up bow.

A simple single acting design used by Martin has an aluminum U channel with a round steel center rod (Fig. 4.19). The center of the steel rod is below the elastic axis of the U channel, so tightening the end nut adds crown.

Single acting truss rods are light and simple, but can only decrease up bow. A partial solution is to tighten the truss rod enough to impose some back bow and level

Fig. 4.18 Cutaway of a nut on a single acting truss rod (photo reproduced courtesy of Taylor Guitars, taylorguitars.com)

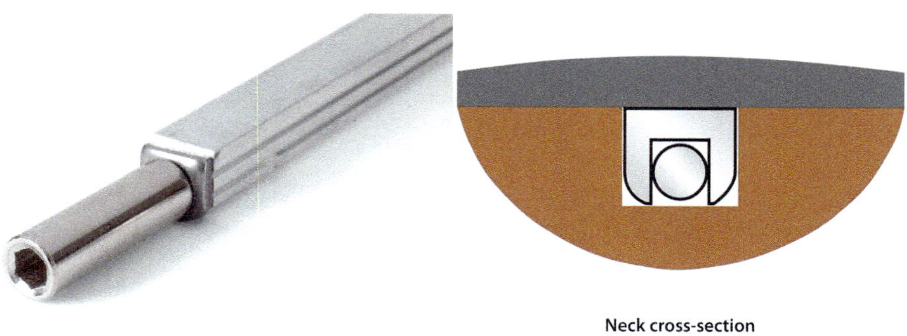

Fig. 4.19 A U channel truss rod (image courtesy of Stewart MacDonald, stewmac.com)

the fretboard. When the truss rod is loosened, the fretboard will have some up bow sanded in. A more general solution is to use a double acting truss rod that can add either back bow (crown) or up bow as needed. While offering more adjustment range, they are heavier and more expensive.

Double acting truss rods are actually made from two rods, one in compression and one in tension. A popular design is called a Hot Rod, and is made of two round

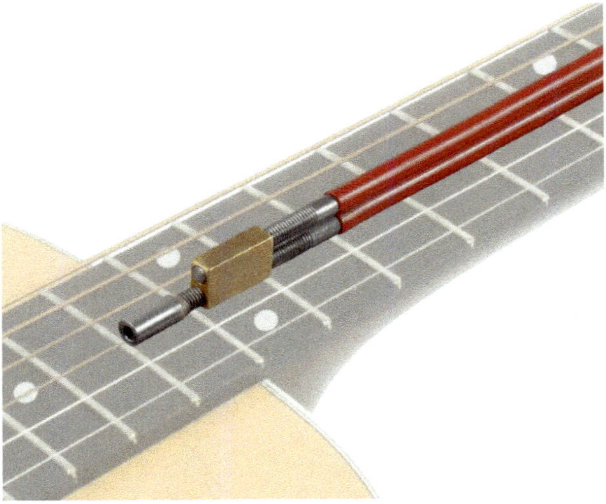

Fig. 4.20 A Hot Rod double acting truss rod (image courtesy of Stewart MacDonald, stewmac.com)

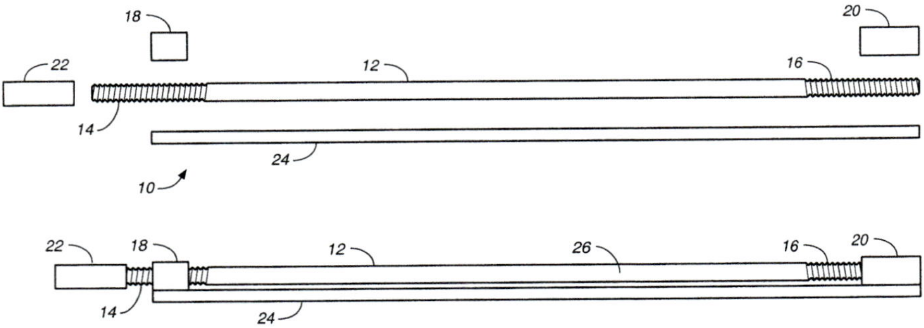

Fig. 4.21 Drawing of a double acting truss rod (Patent 6,259,008, uspto.gov)

threaded rods set into brass blocks at each end (Fig. 4.20). This design is narrow (0.218 in. or 5.54 mm), but tall (0.437 in or 19.05 mm) and can be difficult to fit into some thin necks.

An alternative is low profile truss rod in which one bar is a round threaded rod and the other is a rectangular bar that is welded to the end blocks (Fig. 4.21).

These and other designs work the same way and can induce neck curvature in either direction. Figure 4.22 shows two identical low-profile double acting truss rods adjusted to curve in opposite directions. Both are covered with red heat shrink tubing.

It's important to note that, while very useful, truss rods aren't magic. It is difficult to induce large deformations in a beam while only using forces from the inside (internal forces). There just isn't enough leverage to cause large deflections. Fortunately, large deflections aren't needed. For context, changing the curvature in a neck by 1 mm (0.039 in.) by adjusting the truss rod is rather a lot.

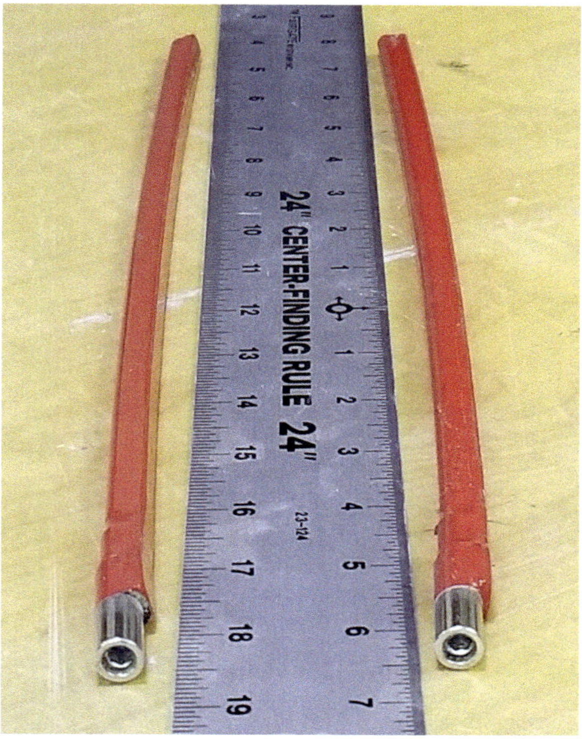

Fig. 4.22 Two identical double acting truss rods showing opposite curvature

Truss rods can be installed with the adjusting nut on either end of the neck. Some manufacturers prefer to have the nut at the headstock, so the design must account for all the material that must be removed to provide access to the nut. Figure 4.23 shows a headstock drilled for access to the adjustment nut of a double acting truss rod.

Others prefer the opposite orientation, with the adjustment nut at the heel. Often, the nut is accessed through a hole drilled through the first transverse brace so all adjustments are through the soundhole. Figure 4.24 shows a truss rod accessible through the soundhole. The adjustment nut is set into a hole drilled through the transverse brace.

Modern classical guitars are sometimes fitted with the same kinds of truss rods used in steel string guitars. An example is the Cordoba C12, shown in Fig. 4.25, that has a double acting truss rod.

The next level of complexity is a multiple rod system that combines fixed and adjustable rods in the same neck. The fixed rods have two purposes, one is to increase neck stiffness to resist bending due to string tension and the other is to increase the resonant frequencies of the neck, reducing instances of "dead spots" on the neck. Dead spots are points along the neck at which string motion from a note fretted there is highly damped and does not ring well. The adjustable truss rod adds additional mass and stiffness, while allowing the relief to be set as desired. The

Fig. 4.23 A double acting truss rod accessed from the neck

Fig. 4.24 Interior truss rod adjustment nut (Wikimedia Commons, commons.wikimedia.org, uploaded by user Gothick)

Fig. 4.25 A Cordoba C12 classical guitar (image courtesy of Guitar Salon International, guitarsalon.com)

Fig. 4.26 An acoustic guitar neck under construction

range of adjustment is smaller since the neck is much stiffer than one without the fixed truss rods, but the fixed rods decrease the amount of adjustment necessary.

Figure 4.26 shows an acoustic guitar neck under construction. The double acting truss rod is in place on the centerline of the neck and with the adjustment nut protruding from the heel. The empty slots on either side of it are for solid graphite fixed rods.

A more integrated approach to the fixed-adjustable truss rod was patented by Charles W. Kaman II in 1979. Colloquially called the Kaman Bar, it combines a substantial fixed aluminum stiffener with a single acting truss rod (Fig. 4.27). This bar results in a very rigid neck, though it requires an elaborate pocket be cut into the neck blank—a reasonable requirement for a factory, but more difficult for individual luthiers.

Most players prefer a small amount of curvature in the neck that is usually called relief. Relief is the distance between a straight neck and one with up bow. Most guitar setups include some relief, which on an acoustic guitar is typically around 0.002–0.010 in. (0.05–0.25 mm) at the eighth fret. Relief is controlled with an adjustable truss rod, but older guitars with fixed truss rods either had the relief sanded into the fretboard or relied on a bit of natural curvature due to string tension.

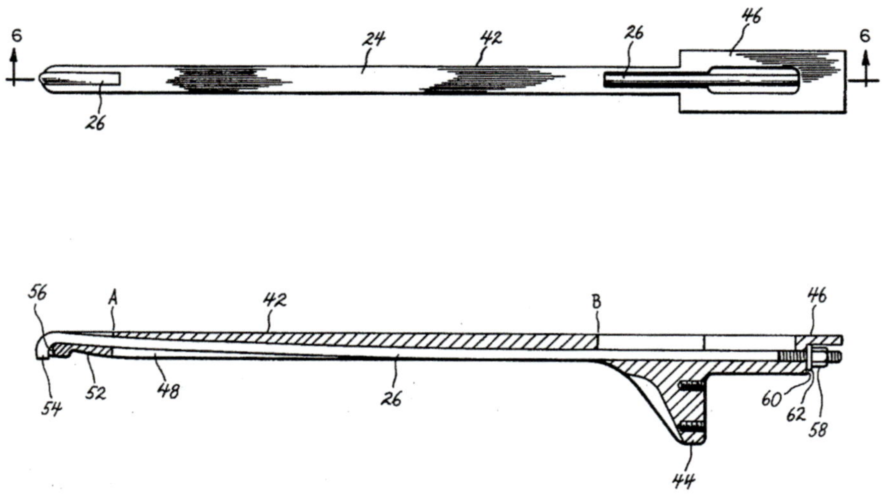

Fig. 4.27 Patent drawing for the Kaman Bar Truss Rod System, Patent 4,172,405

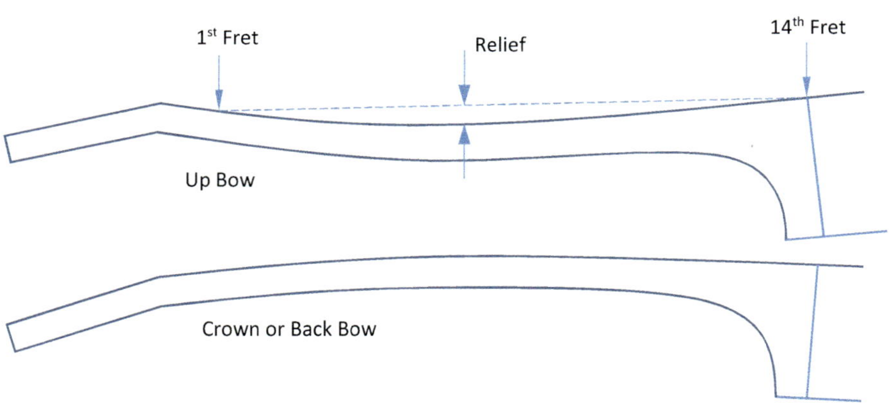

Fig. 4.28 Back bow and crown on a guitar neck

This is particularly true of classical guitars in which necks seldom had an adjustable truss rod. Adjustable truss rods are still far from universal in classical guitars.

Relief can be measured in different ways, but a convenient one is to press a string against both the 1st fret and 14th fret, where the neck joins the body. The string is straight, so the distance between the bottom of the string to the top of a fret is the relief. It's wise to check the relief at more than one fret since the goal is to find the maximum deflection and we don't know exactly where that will be. The neck is tapered so the stiffness varies along its length. There is no reason to think maximum relief will be exactly in the middle of the space between the 1st and 14th frets (near the 6th fret). Fig. 4.28 shows curvature and relief, exaggerated for clarity.

4.2.5 Neck Shape

The cross-sectional shape of the neck is particularly important to most players and many have strong opinions about it. This makes sense, as the neck is the place where players have the most important contact with the instrument. The range of neck shapes now available seems nearly endless, but they can be divided roughly into three classes: U shape, C shape, and V shape (Fig. 4.29).

For initial design, neck shape is important because it defines how much interior volume there is for packaging truss rods and how much material will be in the neck. The final shape of the neck can be refined later in the design process as long as it doesn't cut into the volume needed for truss rods. Individual builders sometimes shape the neck as nearly the final step in construction. Figure 4.30 shows a flamenco guitar in the process of having bindings installed that is structurally almost complete, but with an unshaped neck. This guitar has no truss rod.

Luthiers are constantly experimenting with new neck shapes. An interesting departure from traditional ones is the EndurNeck™ from .strandberg* Guitars (EndurNeck is a trademark of Strandberg Guitars AB and has been patented internationally). It has a trapezoidal cross section with the back flat face not parallel to

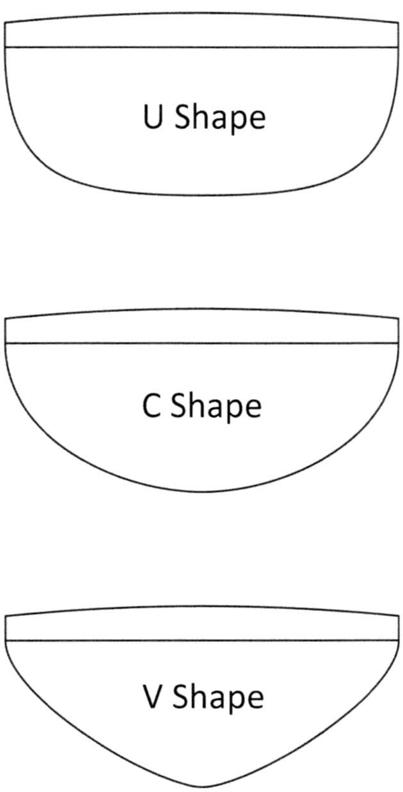

Fig. 4.29 Three basic neck shapes

Fig. 4.30 Flamenco guitar with unshaped neck

the centerline of the neck (Fig. 4.31). I found it to be surprisingly comfortable during a short evaluation. Though it is currently only available on sold body electric guitars, it could be used on acoustic guitars.

While they often have strong feelings about neck shape, musicians don't tend to talk much about neck dimensions other than the width of the nut. However, the taper of the neck is an important part of the basic design. Since necks have a straight taper, it's enough to define neck width and neck thickness in two places. These four dimensions establish neck width and thickness between the nut and heel. The location of the outer strings is defined by the distance from the edge of the fretboard (string setback), as shown in Fig. 4.32. String setback is measured from the outside of the string to the edges of the neck and is usually constant going down the neck.

Table 4.3 gives a representative range of neck dimensions. Note how small the ranges are. One thing the new luthier or guitar designer must get used to is that small changes in some dimensions are important. For example, a very narrow string nut on a steel string guitar is 1–5/8 in. (1.625 in. or 41.3 mm). Conversely, a very wide one is 1–7/8 in. (1.875 in. or 47.6 mm). Most fall well inside that range with the difference in nut widths on production steel string guitars seldom varying more than 1/8 in. (3.2 mm). Classical guitars almost always have wider necks than do steel string instruments.

Fig. 4.31 A faceted neck on a headless electric guitar (image courtesy of . strandberg* Guitars, strandbergguitars.com)

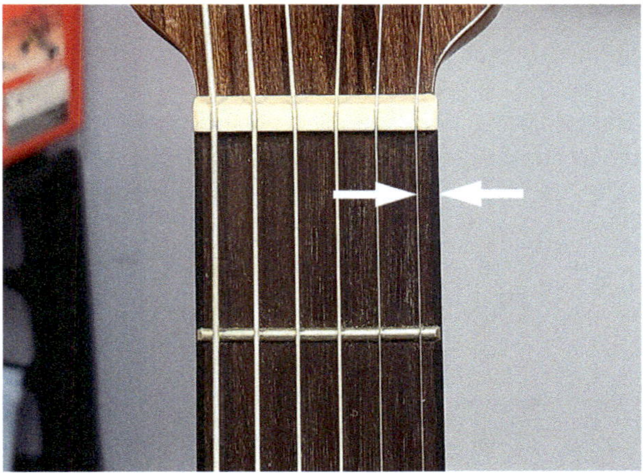

Fig. 4.32 String setback

Table 4.3 Typical range of neck dimensions

Dimension	Classical (in.)	Classical (mm)	Steel String and Archtop (in.)	Steel String and Archtop (mm)
Nut width	1.89–2.13	48–54	1.688–1.75 (1 11/16–1 ¾)	42.9–44.5
Neck width at 12th fret	2.36–2.48	60–63	2.063–2.125 (2 1/16–2 3/16)	52.4–55.6
Neck thickness at first fret	0.75–0.91	19–23	0.813–0.875 (13/16–7/8)	20.6–22.2
String spacing at saddle	2.20–2.36	56–60	2.063–2.375 (2 1/16–2 3/8)	52.4–60.3

Fig. 4.33 A Fender CD-60S dreadnaught guitar (image courtesy of Fender Music, fender.com)

It's important to note that the distance from the outer strings (1st and 6th) to the edge of the neck is approximately constant along the neck on most guitars steel string guitars. Figure 4.33 clearly shows the string spacing on a Fender CD-60S dreadnaught guitar. Distance from the string to the edge of the neck is typically about 1/8 in. (3.2 mm).

It's useful at this stage to write out the width of the neck as a simple equation. First, let's define a couple of dimensions, as shown in Fig. 4.34.

Mathematically, the width has the same form as the straight line we learned in school (y-mx+b).

$$w(x) = \frac{w_s - w_n}{L} x + w_n = \frac{(s_s + 2d) - w_n}{L} x + w_n$$

This expression works in any consistent set of units—inches and mm are both fine.

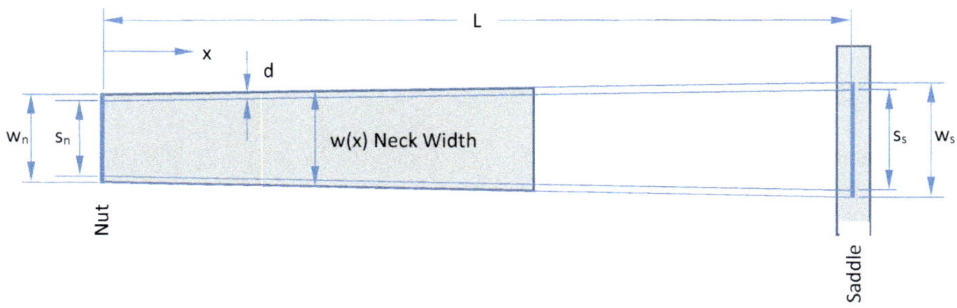

Fig. 4.34 Neck and string geometry

Sometimes, it's more useful to have this expression in terms of fret number, *i*, rather than position along the neck, *x*. The position of fret *i*, measured from the nut is

$$x_i = L\left(1 - \frac{1}{r^i}\right) \text{ where } r = \sqrt[12]{2} \approx 1.05946$$

The width of the neck at fret i is then

$$w_i = \left(s_s + 2d - w_n\right)\left(1 - \frac{1}{r^i}\right) + w_n$$

An algebraic expression like this is compact and unambiguous, but it might help to have some examples. Whether it should or not, the American guitar industry still works in decimal inches. Let's run the numbers in inches for a full-size steel stringed guitar:

- Width of the nut: $w_n = 1.75$ in.
- String spacing at saddle: $s_s = 2.125$ in.
- Scale length: $L = 25.5$ in.
- String setback: $d = 0.125$ in. (note that this is the distance to the center of the string).

Figure 4.35 shows a sample calculation, in inches, for a full-sized steel string guitar, such as a Martin OM.

Figure 4.36 shows a sample calculation, in mm, for a full-sized classical guitar.

For simplicity, these calculations assumed the setback, d, is defined from the edge of the fretboard to the center of the first and sixth strings. Luthiers usually measure setback from the edge of the fretboard to the outer edges of the strings, you may need to offset the calculated value by half the string width.

Modern classical (Spanish) guitars sometimes have additional setback on the first string, especially at higher frets. This helps avoid the string slipping off the edge of the neck. The nut slot for the first string is also often set back more than that for the sixth string. Figure 4.37 shows this in exaggerated form.

$w_n := 1.75$ Nut Width

$r := 2^{\frac{1}{12}} = 1.059$

$s_s := 2.125$ String Spacing at Saddle

$L := 25.5$ Scale Length

$d := 0.125$ Distance from String to Edge of Neck

$$w(x) := \frac{s_s + 2\ d - w_n}{L} \cdot x + w_n$$

$w(0) = 1.75$ Width at nut (check)

$w(12.75) = 2.063$ Width at midpoint (12th fret)

$w(25.5) = 2.375$ Projected Width of Neck at Saddle, w_s

$$w(i) := \left(s_s + 2\ d - w_n\right) \cdot \left(1 - \frac{1}{r^i}\right) + w_n$$

$w(0) = 1.75$ Width at nut (check)

$w(8) = 1.981$ Width at 8th Fret

$w(12) = 2.063$ Width at 12th fret

Fig. 4.35 Neck taper calculations in inches

4.2.6 *Body Shape*

The conventional "hourglass" shape of acoustic guitars has been used for centuries and was in its current form more than 100 years ago. It is familiar, comfortable and clearly works, so it's the obvious choice. However, designers are free, especially at the early stages, to consider different shapes. Centuries of stringed instrument design have produced a wide range of successful instrument body shapes on which to draw.

One problem with the traditional body shape is that it is difficult to build, especially when the top and back radius are introduced. The most basic solution to the problem is a cigar box guitar, a folk instrument whose body is made from the rectangular box in which cigars are shipped. At this writing, cigar box guitars are

$w_n := 52$ Nut Width

$r := 2^{\frac{1}{12}} = 1.059$

$s_s := 60$ String Spacing at Saddle

$L := 650$ Scale Length

$d := 3$ Distance from String to Edge of Neck

$$w(x) := \frac{s_s + 2\,d - w_n}{L} \cdot x + w_n$$

$w(0) = 52$ Width at nut (check)

$w(325) = 59$ Width at midpoint (12th fret)

$w(650) = 66$ Projected Width of Neck at Saddle, w_s

$$w(i) := (s_s + 2\,d - w_n) \cdot \left(1 - \frac{1}{r^i}\right) + w_n$$

$w(0) = 52$ Width at nut (check)

$w(8) = 57.181$ Width at 8th Fret

$w(12) = 59$ Width at 12th fret

Fig. 4.36 Neck taper calculations in millimeters

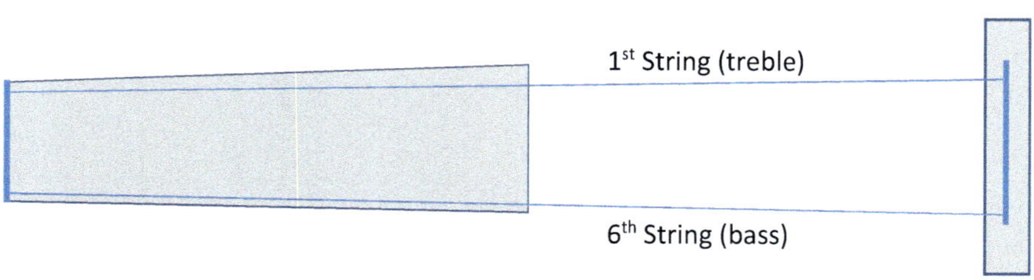

Fig. 4.37 Offset string placement on a classical guitar

experiencing a bit of a resurgence. Figure 4.38 shows one with three strings. As a folk instrument, there are many variations. Note that this one has a soundboard pickup, with the jack visible at the bottom of the instrument (the black surface).

There have been some experimental acoustic guitars with simplified body shapes, but many point back to a simplified violin design proposed around 1819 by Felix Savart, a French physicist and mathematician who was interested in acoustics and structural vibrations. Figure 4.39 shows a drawing of the Savart violin from a

Fig. 4.38 A cigar box guitar with soundboard pickup (image courtesy of C.B. Gitty, cbgitty.com)

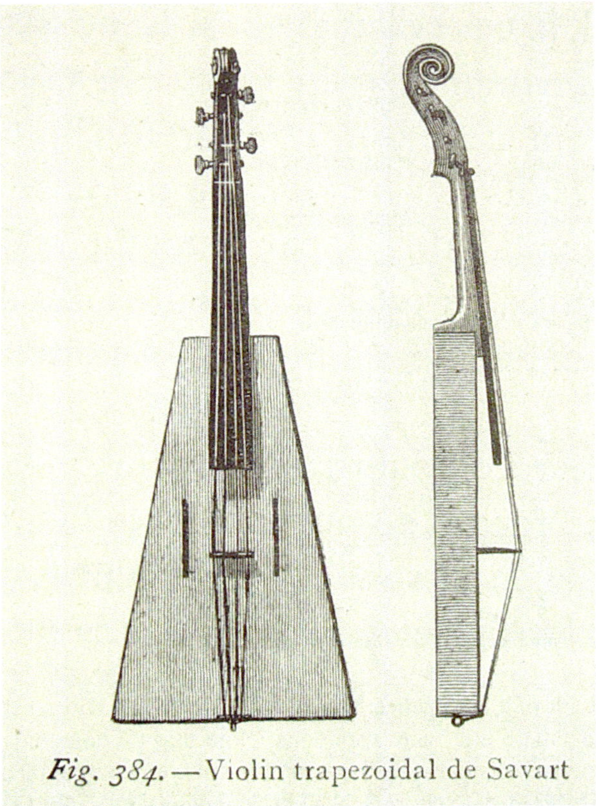

Fig. 4.39 The Savart violin (Wikimedia Commons, commons.wikimedia.org, Biblioteca de la Facultad de Derecho y Ciencias del Trabajo Universidad de Sevilla)

Fig. 4.40 A trapezoidal guitar in Plaquemines Parish Louisiana, 1935 (Original photo by Ben Shahn for the US Resettlement Administration)

Spanish book published in 1882. Though not common, violins have been made on this pattern since Savart proposed it.

Avoiding curved, bent sides makes construction simpler. Indeed, bending sides is sometimes the thing that frustrates beginning luthiers. It makes sense, then, that amateur guitar makers or manufacturers eager to reduce costs have sometimes adopted trapezoidal bodies. Figure 4.40 shows three young women in 1935 with a trapezoidal body guitar. The instrument has wood tuning pegs and a floating bridge.

A few more modern guitar makers have experimented with trapezoidal bodies. Figure 4.41 shows a picture distributed by Martin Guitars of trapezoidal instrument made in 1970, which they describe as the inspiration for the very first guitar made by Chris Martin (C.F. Martin IV).

At the other end of the creative spectrum are builders who have used the traditional body shape as a starting point for further developments. Michael Sankey is a luthier who has experimented with just such refinements. Sankey prefers to give his instruments names rather than model numbers and Fig. 4.42 shows a guitar he calls Breaking Wave.

Few manufacturers have strayed far from traditional body shapes. One rare example from the 1930s is the KayKraft Style A. Built by Stromberg-Voisinet, it was reminiscent of their Venetian mandolin. It has a pressed, rather than a carved arched top and back (Fig. 4.43).

Fig. 4.41 A 1970 trapezoid custom guitar (image courtesy of C.F. Martin & Co. archives, martinguitar.com)

Whatever body shape the designer chooses, the laws of physics still apply. The three most sonically important basic parameters when designing a guitar body are:

1. Body Volume.
2. Soundboard Surface Area.
3. Soundhole Area.

First, the body must have the necessary interior volume. There are details dictated by the complicated physics of structural acoustics, but generally, the lower the frequency the instrument must make, the larger the body volume should be. A

Fig. 4.42 A Sankey Breaking Wave model headless archtop guitar (image courtesy of Michael Sankey, sankeyguitars.com)

Fig. 4.43 A 1930s KayKraft Venetian Style A archtop (image courtesy of Mike and Mike's Guitar Bar, mmguitarbar.com)

guideline is that one octave lower requires 4× the body volume. The body encloses a volume of air that has its own resonant frequency. That frequency is conditioned by the soundhole, which acts approximately like the port on a speaker, and by the flexibility of the top and back plates.

The soundboard radiates much of the sound from the guitar including almost all of the high frequency sound. Thus, it needs enough flexible surface area to radiate that sound. In most guitars, the portion of the soundboard above the lower bout is essentially rigid and doesn't radiate much sound. The flexible lower bout vibrates in response to the strings and acts something like the cone of a speaker.

The other basic dimension is the area of the soundhole. Some people intuitively assume the soundhole "let's the sound out" from the interior of the guitar. This is partially correct, but the reality is more involved. The soundhole does indeed radiate at low frequencies, but its other role is as a port that changes the dynamic response of the soundboard. The area of the soundhole primarily affects the lower frequencies of the body.

To find the volume of a guitar body, it's important to have a mathematical description of the shape. There are several ways to define the shape. The simplest is to draw the body full scale on a large sheet of graph paper. Finding the area requires counting boxes—tedious, but not very hard. A better approach is to model the body in CAD software and have it calculate the area. A rarely used, but mathematically tidier approach is to define the body shape with a mathematical function and just evaluate that whenever you need coordinates. Integrating the function (admittedly, not a basic mathematical operation) gives surface area.

With a shape clearly defined, a family of instruments can be defined by simply scaling it in the X and Y directions to make bodies of different sizes. Figure 4.44 shows how this works. The large blue shape is based on a description by Bob Benedetto of the body shape of a large archtop guitar. It was then scaled to produce the shape in red that corresponds to an OM sized guitar and the shaping in pink for a ¾ sized guitar.

Scaling factors in the X and Y directions can be chosen to produce any body size and this approach has produced instruments down to the size of ukuleles. As any experienced luthier knows, it's important to choose a body shape that fits into commercially available cases and gig bags.

As shown in Table 4.4, if you know the area enclosed by the outline of a guitar body and then scale it in the X direction with a scale factor, S_x, and in the Y direction by a scale factor, S_y, the new area is

$$A_{\text{scaled}} = S_x S_y A_{\text{original}}$$

It's wise to have clear, unambiguous mathematical descriptions of your designs so you can reproduce them whenever needed. It also lets you communicate a precise shape to others. The reason may be something as simple as having cases made or as complicated as working with a supplier to make production forms.

Body volume of a full-sized acoustic guitar is the area enclosed by the body multiplied by its average depth. The total area enclosed by the red curve in Fig. 4.44, similar to an OM, is 222 in^2 (0.143 m^2) and the average body depth of an OM is about 3.8 in. (96.5 mm). However, these are exterior dimensions. The interior of the guitar is slightly smaller because of the thickness of the top, back, and sides and the

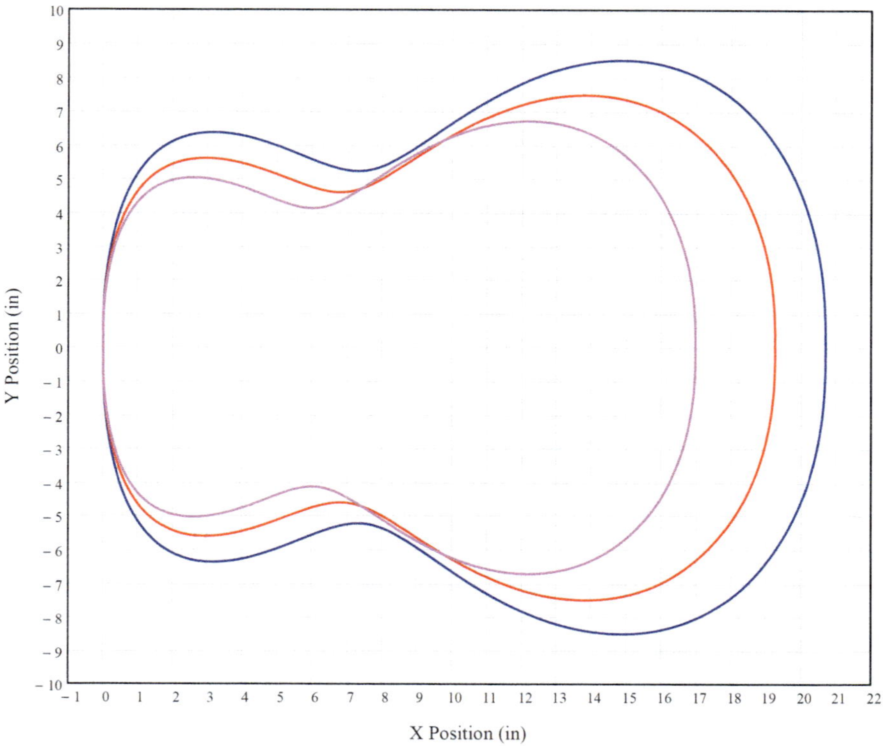

Fig. 4.44 Scaling guitar body shapes

Table 4.4 Body scaling factors

Body type	Body length	Lower bout width	S_x	S_y
Benedetto/Jumbo	20.69 in. 525.5 mm	17 431.8 mm	1.00	1.00
OM/OOO	19.25 in. 489 mm	15 381.0 mm	0.93	0.88
3/4	17 in. 431.8 mm	13.44 in. 341.4 mm	0.82	0.79

volume taken up by the braces, neck block, and tail block. We might assume the total decrease is about 15%. If so, the interior volume is about 11.75 L (717 in.^{3}).

In laying out a guitar body, it helps to have proportions of some representative instrument. Figure 4.45 shows the proportions of the 1933 Martin OM presented in a previous chapter. The image is of a modern OM-21. All numbers are fractions of the body length so these relationships can be applied to any size guitar. Note that these proportions and those that follow are based on instruments that were laid out using traditional drafting methods.

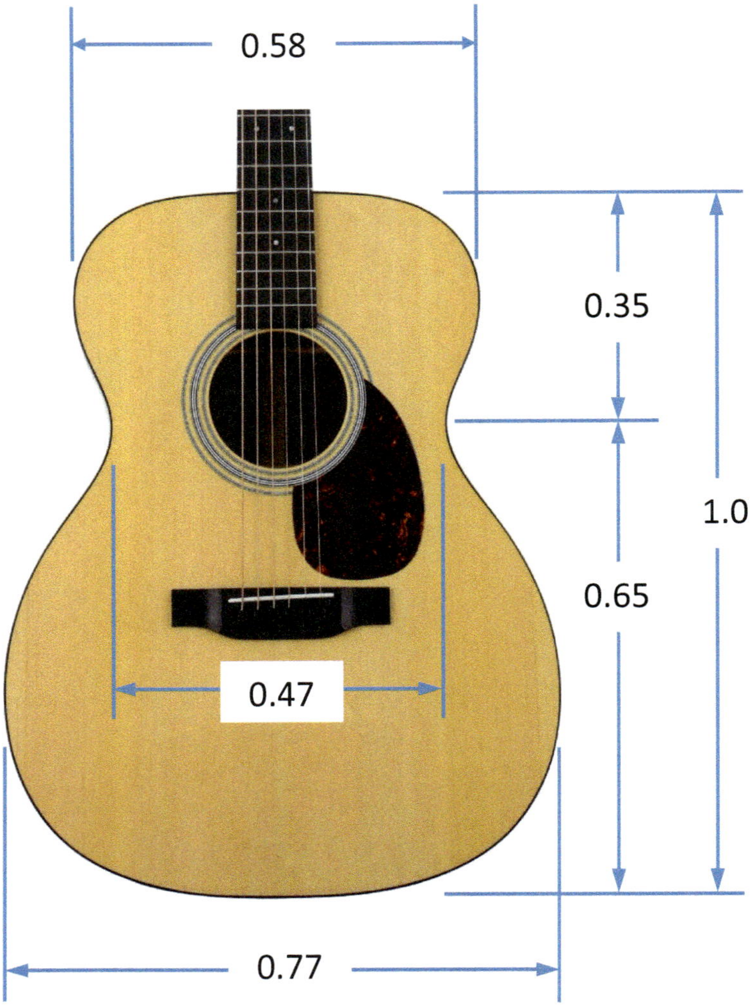

Fig. 4.45 Proportions of a Martin OM guitar (image courtesy of Sweetwater Sound, sweetwater.com)

Classical guitars traditionally have a slightly different body shape than steel string ones. Classical guitars are similar to violins in that they are sometimes made in ¼, ½, ¾, and 4/4 (full) sizes. These numbers are not precise scale factors, but notional sizes.

Formal standards are universal in engineering design and manufacturing, but almost unknown the guitar world. Accordingly, there is no definition of the different guitar sizes, and they vary between manufacturers. Table 4.5 shows representative dimensions for Protégé C1M series guitars by Cordoba (from the Cordoba web site, cordobaguitars.com). This is a series of inexpensive instruments intended for

Table 4.5 Representative classical guitar sizes

Dimension	¼ Size	½ Size	¾ Size	Full Size
Nut width	45 mm/1.77 in.	48 mm/1.89 in.	48 mm/1.89 in.	52 mm/2.05 in.
Upper bout	216 mm/8.50 in.	254 mm/10.0 in.	260 mm/10.2 in.	286 mm/11.26 in.
Lower bout	289 mm/11.38 in.	337 mm/13.27 in.	340 mm/13.4 in.	371 mm/14.60 in.
Waist width	197 mm/7.76 in.	222 mm/8.74 in.	219 mm/ 8.62 in.	241 mm/9.49 in.
Body length	381 mm/15.0 in.	425 mm/16.73 in.	457 mm/17.99 in.	489 mm/19.25 in.
Scale length	480 mm/18.9 in.	580 mm/22.83 in.	615 mm/24.21 in.	650 mm/25.59 in.
Depth (upper bout)	75 mm/2.95 in.	83 mm/3.27 in.	80 mm/3.15 in.	95 mm/3.74 in.
Depth (lower bout)	80 mm/3.15 in.	89 mm/3.50 in.	85 mm/3.35 in.	100 mm/3.94 in.
Soundhole diameter	84 mm/3.31 in.	84 mm/3.31 in.	84 mm/3.31 in.	84 mm/3.31 in.
Overall length	806 mm/31.73 in.	895 mm/35.24 in.	953 mm/37.5 in.	997 mm/39.25 in.

beginning students. Yamaha offers a School series of classical guitars whose ½, ¾, and full-size instruments have different dimensions than those from the Cordoba. The Suzuki program also had instruments of various sizes for children, but I was unable to find dimensions.

A general description of body proportions is by Richard Bruné and shown in Fig. 4.46. Body length is typically around 75% of the scale length. The guitar was made by Kathrin Hauser in 2013.

The D'Angelico New Yorker is about as large as guitars get. Because the scale length is about the same as any other full-sized guitar, it had to grow disproportionately in width, as shown in Fig. 4.47.

4.2.7 Materials

At this writing, acoustic guitars are almost all still made of wood. Tops of all but the cheapest guitars are made of solid wood, usually spruce or cedar. Mahogany tops have been available since before WWII and are becoming more popular as this is being written. A detailed discussion of top wood appears later. For now, it's enough to note that luthiers have mostly gravitated toward a few species of spruce, along with western red cedar and mahogany. Redwood is also occasionally used for tops. Spruce has a high strength to weight ratio and is still plentiful enough to be available in the needed volumes and at acceptable prices. In the US, Sitka spruce is most common, though a few other species are used.

Any wood with acceptable mechanical properties could be used for acoustic guitar tops. Wood for tops needs to have a high strength to weight ratio, a high stiffness to weight ratio, low damping and sufficient hardness to make a durable instrument.

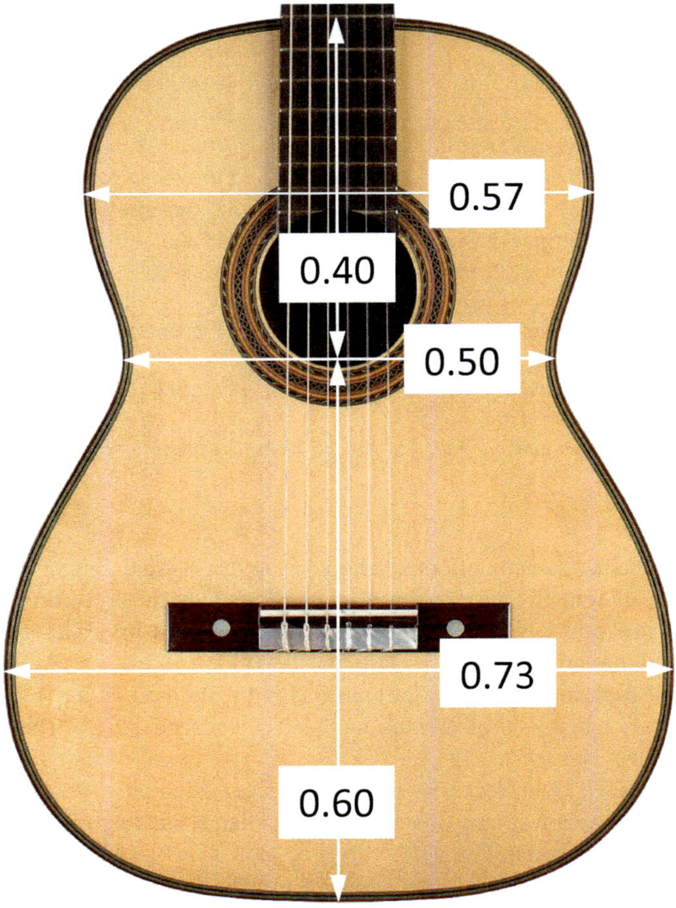

Fig. 4.46 Relative dimensions of classical guitars (after Richard Bruné, rebrune.com, guitar image courtesy of Guitar Salon International, guitarsalon.com)

Table 4.6 lists properties of a few popular species. Note that the mechanical properties of wood vary, even within a single tree, so these numbers are necessarily approximate. Most of these numbers are from the USDA Forest Products Lab. Density is in g/cc which is equivalent to specific gravity. Bending modulus is in GPa and compression strength is in MPa. Bending modulus is just the technical name for the bending stiffness of a material. Side hardness is measured by pressing a steel sphere half its diameter into the surface of the wood. Hardness is reported as the force required to do this. For reference, 4.45 N = 1 lb.

Balsa is strong and light. Based only on that, it would seem to be a good choice for guitars. However, it is soft and easily damaged, so a guitar whose top was made of balsa would have a very short life unless it was always handled carefully. A few luthiers use it for braces, usually with graphite cap strips.

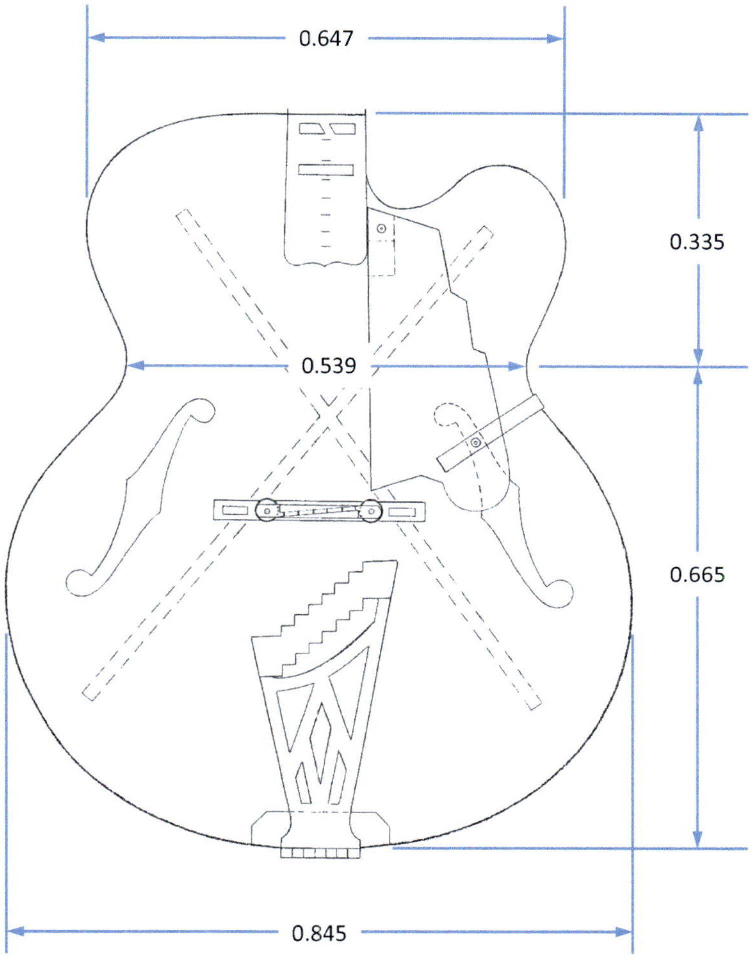

Fig. 4.47 Proportions of the 1948 D'Angelico New Yorker (courtesy of guild of American Luthiers, luth.org, original plan by Steven Andersen)

Table 4.6 Mechanical properties of selected guitar top woods

Wood	Side hardness (N)	Stiffness/weight	Strength/weight
Tropical Balsa	400	21.25	93.13
Ochroma pyramidale			
Sitka Spruce	2300	27.50	99.17
Picea sitchensis			
Western Red Cedar	1600	24.06	98.13
Thuja plicata			
Engelmann Spruce	1750	25.43	88.29
Picea engelmannii			
Old Growth Redwood	2100	23.00	106.00
Sequoia sempervirens			
Honduran Mahogany	3600	17.17	77.83
Swietenia macrophylla			

High material damping causes notes to decay quickly and few luthiers would choose wood with high damping for guitar tops. Woods that otherwise have acceptable mechanical properties are rejected by luthiers because of high damping. Measuring damping accurately can be a difficult, fiddly process, requiring carefully designed supports and attention to mathematical details in post-processing. Damping seldom appears in tables of mechanical properties and would be very approximate if it did. Thus, luthiers learn to tap top plates to listen for notes that decay slowly. Indeed, this may be accurate enough for most builders.

There are several woods that, based on a few commonly reported numbers, should be popular, but aren't. For example, basswood is about as dense as Sitka spruce and has about the same stiffness to weight ratio. It's strength to weight ratio is about that of Engelmann spruce, another popular top wood. Basswood is also popular among wood carvers because it is uniform and easy to work with hand tools. By these measures, it should be a popular choice for acoustic guitar tops, but it is hardly ever used. Hard data is lacking, but one reason is damping. Basswood is more widely used in solid bodies of electric guitars where mechanical properties of the wood have a more minor role in producing tone.

Another design choice is whether to use solid or laminated wood for the top. For almost all individual luthiers and most manufacturers, the only real choice is still solid wood. Laminated wood is durable and can be inexpensive, but the glue layers can change the dynamics of the tops enough to degrade the tonal quality of the instrument. A well-known designer remarked that, with too many seams, "you're not listening to wood, you're listening to glue."

There have been some dedicated efforts to make laminated wood tops that are tonally equivalent to solid ones, but none have yet been satisfactory. It's possible that further experiments will yield a better design that combines the stability of laminated wood with the tonal quality of solid wood tops. A complication is that the ratio of lateral to longitudinal stiffness is quite different for laminated tops, compared to solid ones. This suggests that bracing patterns might need to be developed specifically for laminated tops.

Laminated sides and backs are much more common than laminated tops because they have little effect on the sound quality and are more durable than solid wood. Guitar repair people are routinely presented with cracked sides, so removing this as a cause of failure is valuable. Laminated sides are much less likely to crack, especially when one layer is cross grain. Even builders of fine concert instruments sometimes use laminated sides, though, as you probably expect by now, this idea, too, is hardly a new one. Manufacturers often use laminated sides and backs for lower cost model lines. For example, Taylor uses laminated backs and sides in their popular 100 and 200 series instruments.

A common approach for laminated sides is to glue one or two veneer layers to the inside of a thin, solid wood side. As an example, classical builder John Bogdanovich uses two layers of 0.6 mm (0.022 in.) thick veneer for the inner layers. The outer layer is typically 1.6–1.9 mm (0.065–0.075 in.).

An attractive alternative for the inner layers is aircraft grade birch plywood. This material is approved for load carrying structures on manned aircraft, so the quality

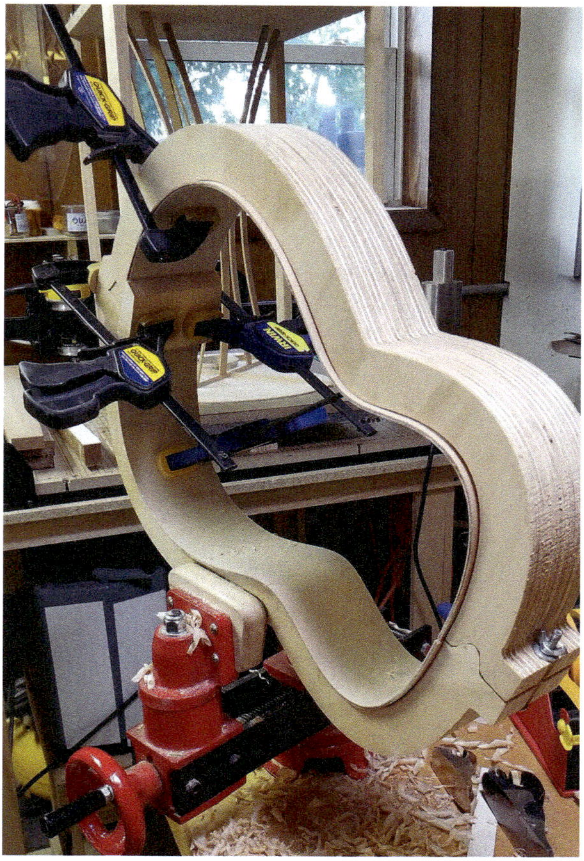

Fig. 4.48 An experimental instrument with laminated sides under construction

and uniformity are as good as wood gets. Aircraft grade ply must not have voids and the glue must pass delamination tests in which samples are immersed in boiling water. It holds up just fine in guitars. It's also available in sheets as thin as 0.4 mm (1/64 in.). 1 mm (3/64 in.), 3 ply material works very well for laminated sides. If the outer plies are perpendicular to the grain of the outer, solid wood, the sides should be stable and impervious to splitting. Figure 4.48 shows the rim of a small guitar being assembled with sides whose inner layer is 1 mm aircraft plywood. The kerfing is not yet installed.

The most common alternative to wood is carbon fiber. A few manufacturers offer carbon fiber instruments, though they are still a small portion of the market. They are also still expensive, though that may to be slowly changing. When some manufacturer is successful enough to produce carbon fiber instruments in large volumes, prices should decrease noticeably. At this writing, it seems likely that the market for carbon fiber instruments will grow, though wood guitars will likely be the norm for the foreseeable future.

Fig. 4.49 Glen Campbell's 1968 Ovation Deluxe Balladeer (image courtesy of the Musical Instrument Museum, Phoenix, Arizona, USA, mim.org, instrument loan courtesy of the Glen Campbell Museum)

Composites, such as carbon fiber, fiberglass, and kevlar, are strong, durable, and largely unaffected by humidity changes, so they are attractive for guitars. They also offer design options that would be difficult to execute in wood. Ovation famously kicked off the market for composite materials in guitars with their Balladeer in the late 1960s, and other manufacturers have slowly entered the market. Fig. 4.49 shows Glen Campbell's prototype Balladeer, with a wood top and composite bowl back.

Figure 4.50 shows a nylon stringed carbon fiber Model X20 instrument by Emerald Guitars, a more recent entrant in the market of carbon fiber guitars.

There are two problems with using composite materials. The first, and probably the less important, is that composite instruments often don't sound quite like wood ones. This matters only if the sound doesn't match players' expectations. If costs decrease, it's easy to imagine that a group of players will emerge who like or even prefer the sound of composite guitars. It's also possible that design refinements will decrease the tonal distance between wood and composite instruments.

The second problem is harder to solve. Composite materials require different manufacturing processes than wood and don't lend themselves conveniently to

Fig. 4.50 An X20 nylon string carbon fiber guitar (image courtesy of Emerald Guitars, emerald-guitars.com)

small shop production. They require precise molds and often require heat to cure the epoxy matrix material. Certainly, anyone willing to do the preparatory work and learn the processes can make composite instruments. However, it is a barrier to entry for luthiers accustomed to wood.

Some manufacturers have introduced other materials, including high pressure laminates (HPL, e.g., Formica) and composites made using linen. At this writing, HPL is the most common. HPL is made from layers of paper bonded with phenolic resin and is often used for kitchen counter tops. It is stable, water resistant, and durable. It is dense, compared to most traditional guitar woods, particularly spruce, so it's initially surprising that it would be used in guitars. However, it has been successful and holds a solid place in the guitar world.

The most notable manufacturer is Martin, which makes a range of instruments from HPL. While some of their HPL instruments have wood tops, others use the HPL for tops, backs, and sides. Figure 4.51 shows a model O-X1E with HPL top, back, and sides. It does use Sitka spruce for the braces. Being a manufactured material, HPL can be made any color and can have pictures or graphic patterns printed on it. This instrument is printed to look like mahogany. Finally, the neck is made from laminated birch and both the bridge and fretboard are of Richlite, another paper-based composite. This instrument should be very tolerant of changes in humidity.

4.3 Electronics

Electronics are less central for acoustic guitars than for electric ones, but many acoustics have some sort of onboard pickup. This is usually accompanied by a preamp with some basic controls and often including a tuner. The range of electronics

Fig. 4.51 A Martin O-X1E guitar made from HPL (image courtesy of Martin Guitars, martinguitars.com)

now available is vast, so this overview focuses only on the most important underlying ideas.

For most of their history, guitars were unamplified for the simple reason that electronics hadn't been invented yet. For context, nobody knew what an electron was until 1897. Lee De Forest developed the first vacuum tube in 1906 (Fig. 4.52) and the transistor wasn't invented until 1947.

A primary reason for amplifying acoustic guitars is performance, especially to large crowds and as part of a band. From the beginning, guitarists playing in bands struggled to be heard among the other instruments. As an example, Fig. 4.53 shows the Benny Goodman Orchestra playing in 1937. Near the center, a guitarist is playing a large archtop. There is no wire coming from it and the only visible microphone is the one into which Goodman is playing. Archtops got larger in an effort to make them louder, but hit a practical limit. The solution was amplification.

There were several early electric guitars, including the Rickenbacker A-22 "frying pan" and the Les Paul's "log." However, the first commercially successful

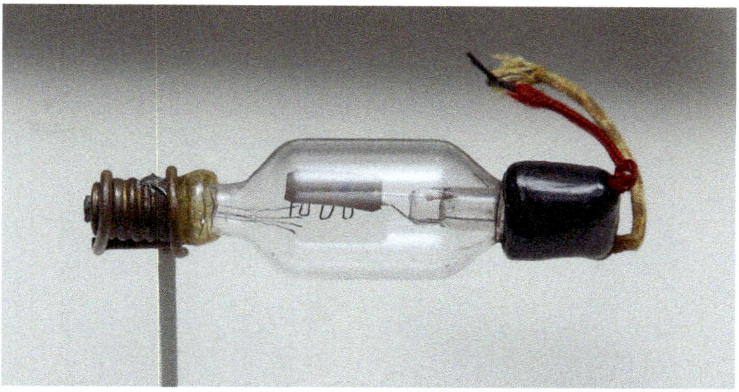

Fig. 4.52 The Audion triode vacuum tube (commons.wikipedia.org, image uploaded by Gregory F. Maxwell)

Fig. 4.53 The Benny Goodman Orchestra in 1937 (image is widely circulated, original source unknown)

electric guitar was the Gibson ES-150, famously played by Charlie Christian (Fig. 4.54).

The ES-150 (ES standing for Electric Spanish) had a single large electromagnetic pickup with heavy ferrite magnets. The three screws visible between the neck pickup and Charlie's right hand secured the pickup to the top.

Fig. 4.54 Charlie Christian playing a Gibson ES-150 in 1939 (Wikimedia Commons, commons.wikimedia.org, original image from Downbeat magazine, photo by Charles B. Nadell)

The field of electromagnetic pickups and electric guitars is enormous and covered well elsewhere. Rather, let's focus here specifically on pickups for acoustic guitars. That is, guitars that still function as acoustic guitars when unplugged and whose tone is largely unaffected by the presence of a pickup and its associated electronics. Pickups for acoustic guitars are mostly of four types: under saddle, soundboard, microphone, and electromagnetic.

Before surveying the different types of pickups, we need to establish the idea of electrical impedance, something central to the operation of many acoustic guitar pickups. Many of us are familiar with the idea of electrical resistance, the property that makes voltage decrease when transmitted down a wire. Resistance is defined in terms of direct current (DC). Impedance is just the equivalent of resistance, but for unsteady or alternating current (AC). Since strings and soundboards vibrate, the output of a pickup is necessarily AC. Impedance of a circuit usually varies with frequency. Resistance is designed by R and impedance is designated by Z. Both Z and R are expressed in Ohms, designated by the capital Greek letter Omega, Ω. 1000 Ω = 1 kiloOhm or 1 kΩ. 1,000,000 Ω = 1 megaOhm or 1 MΩ.

A less intuitive, but important idea is the difference between input and output impedance. Every electronic device has both an input and an output impedance. Roughly, the input impedance, Z_{in}, is the resistance that time-varying signals see when input to a circuit, like an amplifier. Output impedance, Z_{out}, is the resistance that time varying signals see when looking back at the circuit they came from, like a pickup.

Books and articles about electronics, like amplifiers and pickups, often talk about signals. A signal is just a time-varying voltage that means something. Here, the signal is the voltage generated by the pickup. It changes rapidly with time and encodes the vibrations of the instrument. When that signal goes through the amplifier to the speaker, it emerges as music in the form of sound waves.

Generally, the input impedance of an amplifier should be much higher than the output impedance of the pickup. If the ratio is much less than 10:1, the volume is noticeably reduced.

A simple explanation uses resistors to show approximately how it works. Replacing impedance with resistance is correct enough to illustrate the basic idea. Figure 4.55 shows a voltage divider made from two resistors. The pickup provides the input voltage, V_{in}. R_{pu} is the output resistance from the pickup and R_{amp} is the input resistance of the amplifier. The output voltage is the voltage dropped across the amplifier. In circuit terminology, the amplifier is the load, which is often written as R_{load}.

The ratio of output voltage to input voltage, V_{out}/V_{in}, describes what fraction of the signal from the pickup gets to the amplifier. The goal is to send as much of it to the amp as possible. Electrical engineers often express this as dropping as much voltage as possible across the load. The ratio is

$$\frac{V_{out}}{V_{in}} = \frac{R_{amp}}{R_{amp} + R_{pu}}$$

So, if $R_{amp} = 10R_{pu}$, 91% of the voltage from the pickup gets to the amplifier. Conversely, if $R_{pu} = R_{amp}$, only half the voltage from the pickup gets to the amplifier. In practice, this would lower the volume that could be produced by the amplifier since it only sees half the output of the pickup. The pickup will still work when Z_{in} is low, but the volume will be lower and frequency response may be affected.

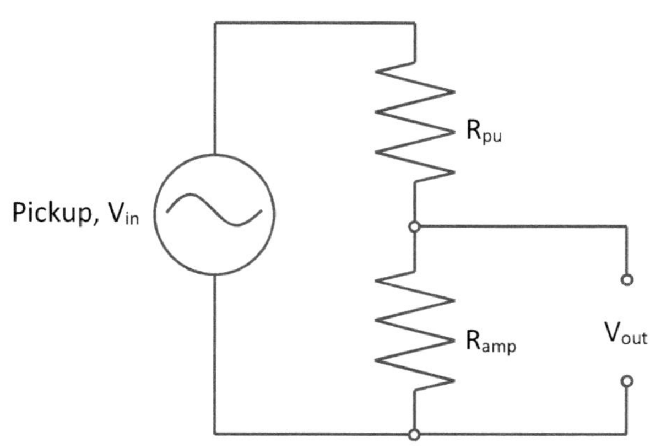

Fig. 4.55 A DC voltage divider

Impedance is a function of frequency, so it's a curve, not a number. However, people often use a single number, the impedance at 1 kHz, as a rough approximation to the function. With this in mind, the output impedance, Z_{out}, of an electric guitar is often in the range of 10–20 kΩ. Thus, the input impedance, Z_{in}, of an amplifier should be more than 200 kΩ. Few amps have Z_{in} of less than 470 kΩ and many are 1 MΩ or more. For example, the manual for the Marshall JVM 410HJS lists the input impedance as 1 MΩ.

If a pickup has high enough output impedance to cause problems, the most common solution is to put a preamp between the instrument and the amplifier. Preamps are designed to have very high input impedance, as much as 10 MΩ, and the output impedance is low, typically less than 1000 Ω or 0.001 MΩ. Gain through preamps is often not high as their main purpose is to eliminate impedance problems.

4.3.1 Under-Saddle Pickups

One of the most common types of acoustic guitar pickups are under-saddle piezoelectric elements. Piezoelectricity was discovered by Pierre and Jacques Curie in 1880. It is the accumulation of charge at the surface of certain solid materials in response to mechanical deformations. Materials that exhibit piezoelectricity include quartz, topaz, cane sugar, and some ceramics. A popular piezoelectric polymer is PVDF (Polyvinylidene Fluoride), sold under several trade names, including Kynar®.

Sensors are devices that turn some physical quantity into a proportional voltage. Piezoelectric elements can make good sensors and are present in a wide range of products, including guitar pickups.

Acoustic guitar pickups sense vibration of some part of the instrument and produce a signal that can be amplified. Under-saddle pickups are thin piezoelectric elements mounted in between the saddle and the bridge, as shown in Fig. 4.56. Because of its placement, the pickup senses the dynamic force between the saddle and the bridge. It also bears a static load from string tension.

Figure 4.57 shows the simplest example of an under-saddle pickup, an AG series passive pickup by Fishman. This model is wired directly to the output jack and may

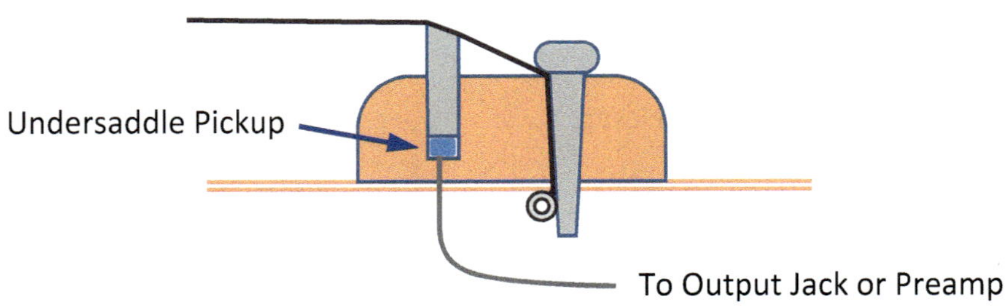

Fig. 4.56 Installation of an under-saddle pickup

Fig. 4.57 Under-saddle piezo pickups (image courtesy of Stewart MacDonald, stewmac.com, pickups by Fishman, fishman.com)

use an external preamp. The different colored coverings indicate different sized pickups.

Pickup manufacturers don't always report output impedance, but the Installation Guide for the Fishman SBT series recommends a preamp with a 10 MΩ input impedance, which hints at an output impedance on the order of 1 MΩ. Accordingly, they offer a range of preamps with Z_{in} = 10 MΩ. They note that connecting to an amp with 1 MΩ input impedance will work, but there will be some roll off at bass frequencies. This is because the lower Z_{in} allows the combination of pickup and amp to act as a filter.

Piezoelectric materials are not conductive, so they develop a charge, rather than a voltage. However, voltage and charge are related.

$$C = \frac{q}{V} \quad \text{or} \quad V = \frac{q}{C}$$

where C is capacitance, q is charge, and V is voltage. Capacitance is a function of geometry and the properties of the material in the pickup. The amplifier needs to see some current from the pickup because of Ohm's law

$$V = iZ_{\text{amp}}$$

where V is the voltage drop across the amp, i is current, and Z_{amp} is the input impedance of the amplifier. Higher input impedance means less current required for a given voltage drop.

Manufacturers of under-saddle pickups typically also offer compact preamps that are easy to mount inside the guitar body. A tidy example is the Element system by L.R. Baggs. The flexible piezo element is mounted under the saddle and the preamp, with integrated jack, is built into the endpin. The volume control is mounted inside the soundhole where it is accessible to the player but hidden from the audience. Figure 4.58 shows a model designed for classical guitars. The piezoelectric element is housed in the braided cable.

Solid piezo ceramic elements are not very flexible and need more careful packaging. A solution is the LB6 from L.R. Baggs in which six piezo elements are mounted in a rigid support to which the saddle is bonded (Fig. 4.59).

An interesting and successful variation on piezoelectric pickups is the ES2 from Taylor Guitars (Fig. 4.60). It uses piezo elements behind the saddle so that the pickup does not bear the string loads and senses the longitudinal dynamic loads between the bridge and saddle. This is a fundamentally different mechanism than an under-saddle pickup.

4.3.2 *Soundboard Pickups*

The movement of the soundboard produces much of the sound from an acoustic guitar, so it makes sense that sensing vibrations directly from the soundboard is a good way to amplify one. It's certainly closer to the mechanism that produces sound than the internal forces sensed by an under-saddle pickup.

Fig. 4.58 An under-saddle pickup with compact preamp (image courtesy of L.R. Baggs, lrbaggs.com)

Fig. 4.59 Unitary saddle and piezo pickup (image courtesy of L.R. Baggs, lrbaggs.com)

Fig. 4.60 Taylor ES2 bridge pickup on a 12-string guitar (image courtesy of Taylor Guitars, taylorguitars.com)

The simplest soundboard pickups are piezoelectric disks of the kind sometimes sold as buzzers or contact microphones (Fig. 4.61). This one has two sensing areas with solder pads. The metal backing disk acts as the ground plane as well as being the mechanical support for the piezo elements. When the plate bends in response to the vibration of the surface it's mounted to, it makes a signal. Several manufacturers offer variations of piezo disk pickups and some have several individual elements (e.g., K&K pickups).

A more sophisticated example of a soundboard pickup is the iBeam from L.R. Baggs shown in Fig. 4.62. It uses piezo film sensing elements bonded to a light plastic beam so that it senses bending. The beam is mounted to the bridge plate, in front of the bridge pin holes, as shown in Fig. 4.62. The battery box is mounted to the neck block for easy access, the volume control is just inside the edge of the soundhole, and the preamp is integrated into the end pin. The installation manual lists its output impedance as 600 Ω, so it can drive any guitar amplifier.

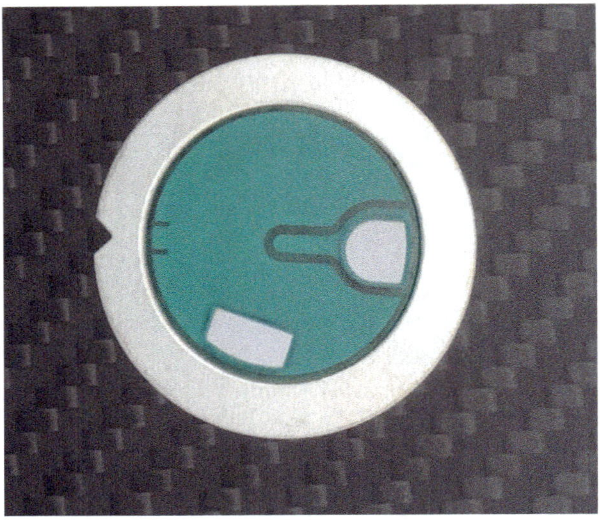

Fig. 4.61 A two element piezo disk

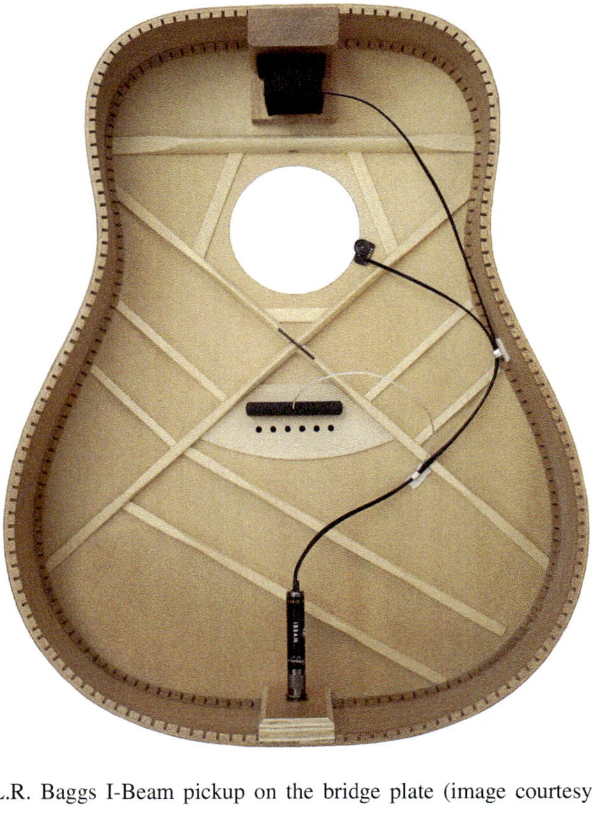

Fig. 4.62 An L.R. Baggs I-Beam pickup on the bridge plate (image courtesy of L.R. Baggs, lrbaggs.com)

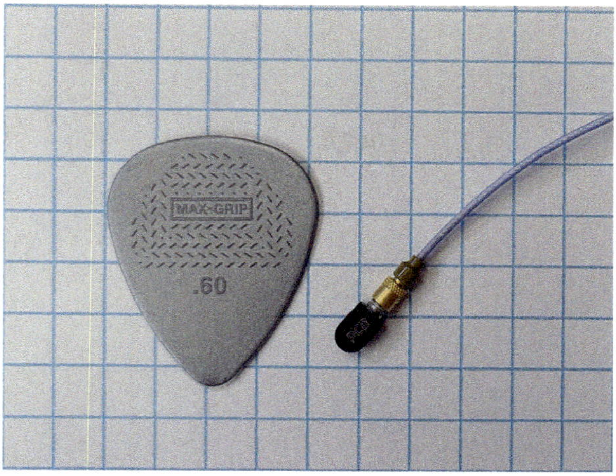

Fig. 4.63 A very small piezoelectric accelerometer

Another type of soundboard pickup directly senses acceleration of the soundboard. Instrumentation accelerometers can do this. They are sensitive, low noise, and very small ones are readily available. However, they are expensive (about the cost of an entry level Taylor guitar) and require specialized signal conditioning. Figure 4.63 shows a PCB model 352C23 accelerometer I use for measuring the dynamic response of acoustic guitars. It weighs 0.007 oz. (0.2 g) so the mass it adds is negligible. It makes a nice soundboard pickup, but is not practical. The grids in this picture are ¼ in (6.35 mm).

An alternative is a soundboard pickup developed by Taylor Guitars as part of their Expression System. Called a Dynamic Body Sensor®, it uses a coil, similar to that on an electromagnetic pickup, to sense the motion of a small mass inside the coil. It is much less expensive than an instrumentation accelerometer, is sensitive enough, and interfaces easily with electronics already used in guitars. Figure 4.64 shows a typical mounting.

Figure 4.65 shows the complete Express System mounted in an acoustic guitar. The soundboard sensor is on the bass side of the lower bout (right side of the picture).

Some other interesting features are also visible. The plate mounted on the bridge plate is a bridge ground. The ball ends of the strings contact it and the coiled ground wire is just visible going to the battery box mounted in the tail block. There is a small electromagnetic pickup under the fretboard which is connected to the preamp by the wire above the transverse brace. The small oval coil on the bass side of the upper bout has no magnets and provides noise cancelling for the electromagnetic pickup.

Fig. 4.64 The Taylor Dynamic Body Sensor® soundboard pickup (image courtesy of Taylor Guitars, taylorguitars.com)

Fig. 4.65 The Taylor Expression System (image courtesy of Taylor Guitars, taylorguitars.com)

4.3.3 Microphones

The simplest way to amplify an acoustic guitar is to put a microphone in front of it, an approach that is routine in recording and in stage performance. Figure 4.66 shows mandolin player Danny Knicely and guitarist James Leva in concert with their instruments amplified by microphones. This works as long as the musicians don't move very much.

Fig. 4.66 James Leva and Danny Knicely, February 2012 (Wikimedia Commons, commons.wikimedia.org, image by Reed George)

Fig. 4.67 A combined microphone and under-saddle system (image courtesy of L.R. Baggs, lrbaggs.com)

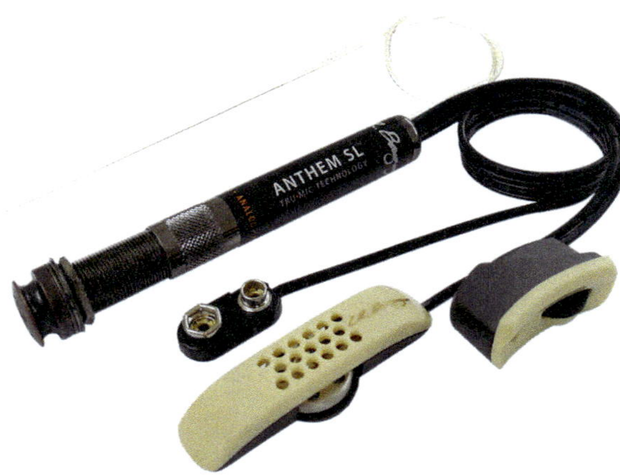

A more flexible approach is to mount a microphone inside the body of the instrument. Though susceptible to feedback, this can work well and interior microphones are sometimes combined with other types of pickups. Figure 4.67 shows the Anthem SL system from L.R. Baggs, which combines a piezoelectric under-saddle pickup and a microphone.

4.3.4 Electromagnetic Pickups

Steel strings are just that. They have steel cores and that allow them to work with the same kind of electromagnetic pickups used in electric guitars. They sense string motion (velocity), so they observe a different phenomenon than the other types.

They operate by electromagnetic inductance—current is induced in a wire that moves with respect to an electromagnetic field. The effect was discovered by Michael Faraday in 1831 and is described by Maxwell's equations. It doesn't matter whether the wire is moving or the field is moving, as long as there is differential motion. Indeed, there is no way the system would know which one was moving.

Electromagnetic pickups consist of a coil of fine wire and one or more magnets. The magnets create a field in which the coil and the string are placed (Fig. 4.68). The obvious question is how the coil senses the string, since the coil is fixed with respect to the magnet. Electromagnetic forces are reciprocal, so the field responds to the string by vibrating slightly as well. It is this vibrating field that the coil senses. The effect is a small one, so it takes many turns of wire to generate a useful signal. Electric guitar pickups can have 4000–8000 turns of very fine wire. Pickup wire is typically 42 AWG or 43 AWG. 42 AWG wire has a diameter of 0.0025 in. or 0.063 mm. 43 AWG wire has a diameter of 0.0022 in. or 0.056 mm.

Taylor is one of the few manufacturers that builds electromagnetic pickups into acoustic guitars. In other brands, it's more common for them to be mounted in the soundhole, almost always as a modification to an existing instrument. Figure 4.69 shows an electromagnetic pickup mounted in the soundhole of an Alvarez guitar.

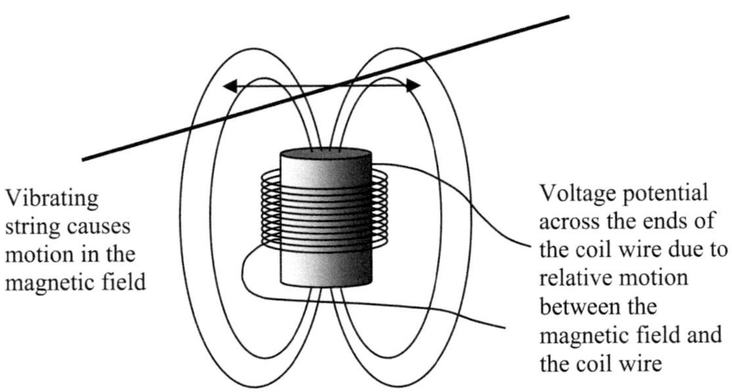

Fig. 4.68 An inductive pickup

Fig. 4.69 An electromagnetic soundhole pickup (Wikimedia Commons, commons.wikimedia.org, image by user beth h)

Bibliography

Books

Bogdanovich, J.S., Classical Guitar Making: A Modern Approach to Traditional Design, 2nd ed., Sterling, 2007, ISBN 978-1402720604

Dailey, Denton J., Electronics for Guitarists, Springer 2012, ISBN 978-1461440864

Fletcher, N.H. and Rossing, T.D., The Physics of Musical Instruments, 2nd ed., Springer, 2010, ISBN 978-1441931207

Forest Products Lab (n.d.), Wood Handbook – Wood as and Engineering Material, FPL-GTR-190, US Department of Agriculture, Forest Service, Forest Products Lab, fpl.fs.fed.us/documnts/fplgtr/fpl_gtr190.pdf

French, R.M., Technology of the Guitar, Springer, 2012, 978-1489987907

Johnston, R. and Boak, D., Martin Guitars: A History, Hal Leonard Books,Milwaukee 2008

Mottola, R.M., Mottola's Cyclopedic Dictionary of Lutherie Terms, 2020, ISBN 978-1-7341256-0-3

Shaw, R. and Szego, P., editors, Inventing the American Guitar, Hal Leonard Books 2013, ISBN 978-1-4584-0576-0

Articles

Bruné, R.E., The Cultural Origins of the Modern Guitar, GFS Soundboard, Fall 1997, pp 9-20.

Caldersmith, G., Designing a Guitar Family, Applied Acoustics Vol. 46, 1995, pp. 3-17

Fliescher, Helmut and Zwicker, Tilmann, Investigating Dead Spots of Electric Guitars, Acustica (Acta Acustica) 85, 1999, pp. 128-135

Freeman, D., Building for Playability – 2008 GAL Convention Lecture, American Lutherie, Vol. 99, Fall 2009, p. 6.

French, M., An Improved Body Shape Definition for Acoustic Guitars, ArXiv pre-print 1802.000439, also Savart Journal, July 2017, savartjournal.org.

Mottola, R.M., Drawing Acoustic Guitar Body Outlines Using Traditional Techniques, American Lutherie, Vol 129, 2017, p. 52

Mottola, R.M., Fretboard Top Plate Geometry of the Flattop Guitar, American Lutherie, 111, 2012, p. 32

Mottola, R.M., Rapid Prototypes for the Flattop Guitar, American Lutherie, Vol. 88, 2006, p. 42., also in Flattop Guitars, p. 68.

Shepherd, M.R., Hambric, S.A. and Wess D.B., The Effects of Wood Variabiity on the Free Vibration of an Acoustic Guitar Top Plate, Journal of the Acoustical Society of America, Vol. 136, No. 5, 2014.

Zoran, A., Welch, S. and Hunt, W.D., A Platform for Manipulation and Examination of the Acoustic Guitar. Applied Acoustics, Vol, 73, No. 4, April 2012, pp. 338-347

Chapter 5
Physics of Acoustic Guitars

Before we can proceed with detail design, it's important to understand the basic physics of acoustic guitars. Design has to be motivated by some goal and the first goal for a guitar designer has to be producing a successful musical instrument. Current understanding of the basic physics of acoustic guitars is good, much better than that of the people who designed the iconic guitars presented in an earlier chapter. However, there is much work to be done on quantifying and predicting sound quality, as well as on robustly linking design changes to improvements in sound quality. Andy Powers notes:

> The guitar serves to remind me that all sciences are merely a way to understand and make sense of what is going on around us. So, when I hear a guitar sound so beautiful it moves me, and I think of all the ways that instrument deviates from the perfection my model would suggest, I'm left to conclude that my understanding, my formula, or my model is simply not yet robust enough to describe everything that is going on.

For an acoustic guitar, the designer must have a working understanding of the physical principles by which a guitar works. Let's start with an understanding of how energy flows through a guitar, from the vibrating strings to the sound waves radiated into the air. Figure 5.1 shows a diagram based on one by Fletcher and Rossing in their book on the physics of musical instruments. Note that low frequency energy and high frequency energy can take different paths. Also, the sides and back are not major sources of radiated sound.

All the energy that appears in the air as sound (pressure waves) came originally from the vibrating strings. If the structure of the guitar was transparent to the dynamic forces from the strings, passing them unmodified, the sound would be just a combination of the fundamental string frequencies and their harmonics and all guitars with the same kind of strings would pretty much sound alike. This isn't what happens, though. The guitar's structure modifies the vibration of the strings and then radiates it as sound. The structure of the instrument acts like something an engineer might call a filter.

R. M. French, *Acoustic Guitar Design*,
https://doi.org/10.1007/978-3-030-89381-1_5

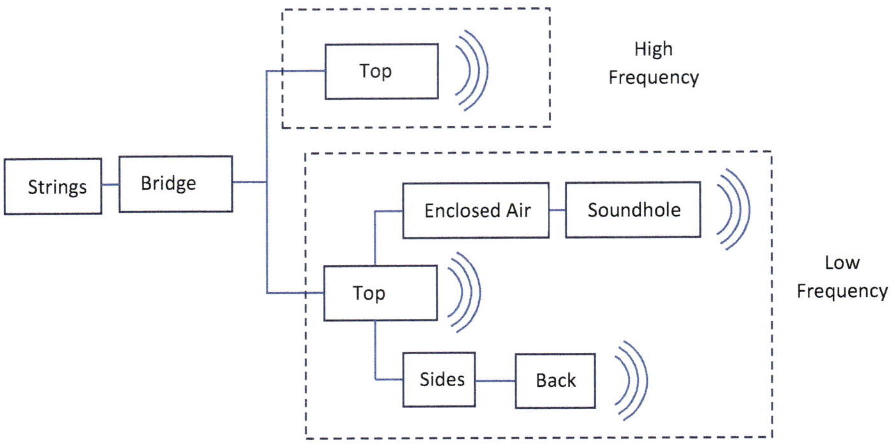

Fig. 5.1 Energy flow through guitar

When we work on the detail design of an instrument and introduce refinements intended to improve the tone, we're trying to adjust how the structure tailors the energy flowing through it and how it radiates that energy out into the air around the instrument.

5.1 Describing Vibrations

A simple way to present the dynamic behavior of the structure is its frequency response function (FRF). The FRF is the response of the structure to a measured input. To collect one, it's common to tap the top with an instrumented hammer that senses force. Top motion is usually measured using a miniature accelerometer, though other kinds of sensors are sometimes used. The ratio of measured accelerometer output over hammer input, mathematically transformed into the frequency domain, is the FRF.

Figure 5.2 shows this kind of tap test. The strings are heavily damped with strips of foam. The response is being recorded using a very small accelerometer placed on the top, next to the bridge and an instrumented hammer tapping the bridge next to the accelerometer. This was a test instrument made by Taylor, who drew the bracing pattern on the top before applying finish. Also, the two brown rectangles labeled 120 Ω and 350 Ω are strain gauges left over from another test. The guitar was supported on soft foam blocks so that it's free to vibrate, without any interaction with the table.

Before we go any further, we need to understand the difference between time domain and frequency domain. I live in time domain and I'd bet that you do, too. I don't know how to live any other way. When we record data, we necessarily record it as a function of time. That's time domain. However, it's often very difficult to

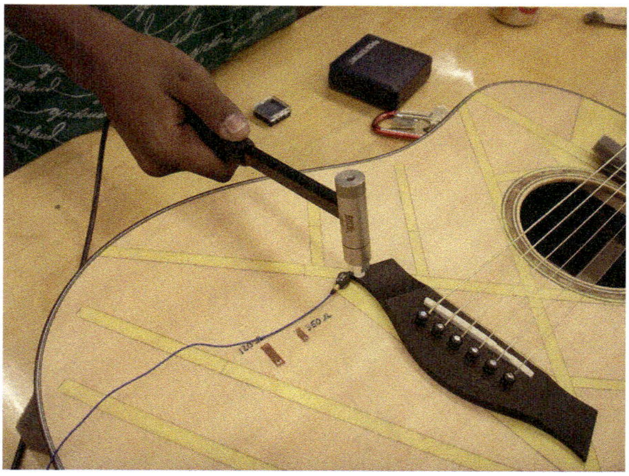

Fig. 5.2 Tap testing a Taylor 710

extract the frequency information we need by looking at the time domain recording. Rather, we need to look at it in some different way. Fortunately, there are common mathematical tools that transform the measured data without changing what information is presented. One of these is the Fourier transform, implemented in a particularly efficient way in signal processing software as the Fast Fourier Transform, or just FFT.

The FFT presents the information recorded by the sensors as a function of frequency rather than time. When implemented correctly, it doesn't destroy any information. The same information that was originally recorded is just presented in a different form that makes it easier to look at and easier to extract meaning.

As a rough analogy, imagine having $20 and converting it to Euros. At this writing $20 = €16.44. Assume there is no fee for the exchange, so nothing is lost. Having $20 in one hand and €16.44 in the other means having the same amount of money in each—the same buying power. They are just in different forms. Buying a set of strings in New York requires dollars. Buying those same strings in Stuttgart requires Euros. Converting one currency to the other is akin to a transform.

The mathematical details of the FFT literally fill textbooks and are interesting, but a distraction here. It's enough to note that a well-measured FRF clearly shows the resonant frequencies of the instrument, based on recordings made by tapping it. Figure 5.3 shows an FRF of the Taylor dreadnaught guitar from Fig. 5.2.

In this plot, the first body resonance is shown as the peak at 99 Hz, the second is at 181 Hz, and the third is at 325 Hz. There is an anti-resonance at 114 Hz and another at 244 Hz. The guitar really doesn't like to vibrate at these frequencies, at least not with the tapping point I used.

It's important to note two things. The first is that the strings were heavily damped with strips of foam so that this plot shows the response of the top and not that of the strings. Without that damping, the accelerometer would have sensed hardly anything but the vibrating strings. The dynamics of the body subtly modify the vibrations from the strings and the goal of this measurement was to measure the dynamics of the body under string tension, but unaffected by string vibration.

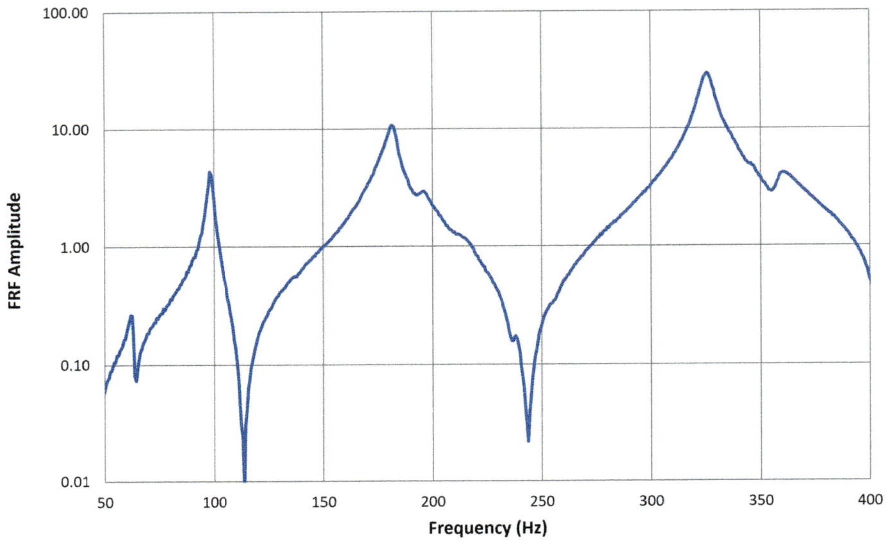

Fig. 5.3 Frequency response function for a Taylor 710

The second important property of FRFs like this one is that it was measured at a single point and only expresses the dynamics of the instrument between that driving point and that response point. Moving the point at which the hammer taps or the place were the accelerometer is placed will give a different FRF. Taking data at a collection of points around the instrument can give complete mode shapes for an instrument, but the testing is tedious and processing it requires sophisticated and expensive software. Sophisticated optical test methods are faster, but much more expensive.

Every structure has frequencies at which it wants to vibrate, called resonant frequencies or natural frequencies. Energy input to the structure naturally wants to come out at these frequencies. Engineers usually refer to the input energy as excitation and talk about exciting structures. The vibrating strings of an acoustic guitar make dynamic forces that excite its structure.

Structures act like filters in that whatever energy goes in, like that from vibrating strings, tries to come out in the resonant frequencies of the guitar. If the strings are damped and the guitar is tapped and allowed to respond freely, the result is dominated by frequencies of the structure. When the undamped strings are plucked and allowed to vibrate, the result is dominated by the resonant frequencies of the strings, but modified by the dynamics of the body and neck. This is central to why luthiers are so interested in the dynamics of their instruments.

Stretched strings are about the simplest possible structures, with resonant frequencies that are easy to calculate approximately. The expression for the frequencies of an ideal string, in Hz, is

$$f_n = \frac{n}{2L}\sqrt{\frac{T}{\rho}} \tag{5.1}$$

where T is tension, L is length, and ρ is mass per unit length. It's a good idea to do this calculation in metric units (kg-m-s) since the mass unit for a lb-ft-s system is the slug and length would have to be expressed in ft. In this equation, n represents which of the frequencies, called partials, you wish to calculate. Structures have many resonant frequencies and ideal strings are unique in that all their resonant frequencies are multiples of the first one, called the fundamental. For the fundamental frequency, $n = 1$. In the music world, frequencies corresponding to $n = 2, 3, 4$, and so on are often called harmonics.

Thus, a string tuned to A at 110 Hz should have harmonics at 220 Hz, 330 Hz, and so on. When a plucked string is allowed to ring, its motion is made up of a combination of these frequencies. There are different relative contributions at each resonant frequency depending on where the string is plucked and on the dynamic properties of the guitar.

Figure 5.4 shows an example of calculated sine waves at 110, 220, and 330 Hz. These correspond to the notes A_2, A_3, and E_4 in the equal tempered scale used by guitars and other fretted instruments. The bottom plot shows a combination of the

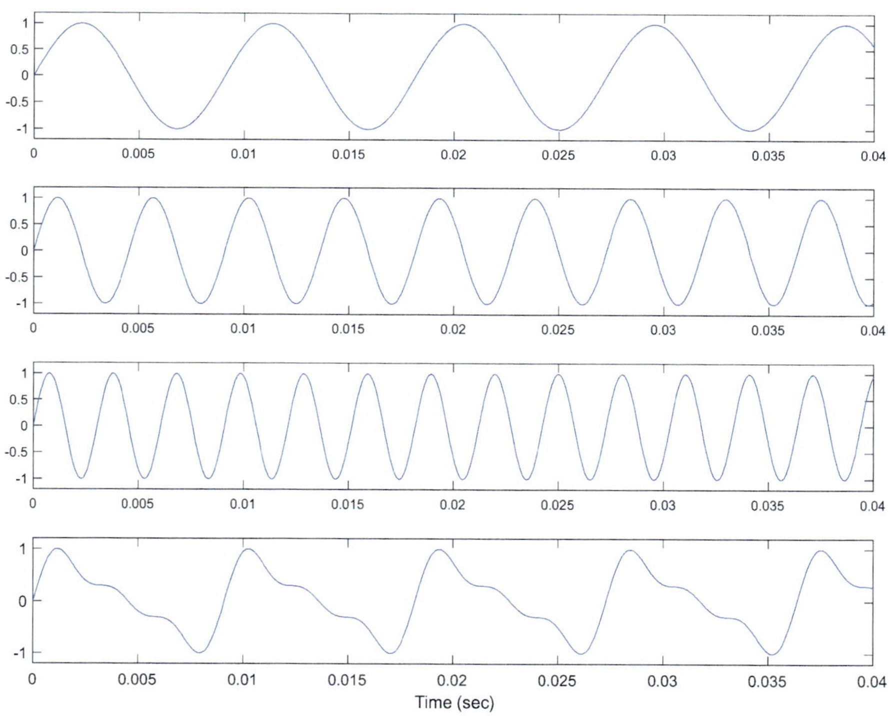

Fig. 5.4 Components of string vibration for an A string

three waves, with differing proportions of each. Even with only three components, it's difficult to see what frequencies are present and in what proportion. A vibrating string might have 20 such components, so we almost always look at response in frequency domain.

Stretched strings are a nice way to make music because they make frequencies that are close to being integer multiples of the fundamental and these correspond nicely to notes in the musical scale. Mathematicians call these successive notes a harmonic series.

Other structures, like vibrating bars, don't make harmonic series. Thus, it's harder to make music with them. If you get a chance, look at the back of a marimba bar. It is carved out in an effort to tune its first few resonant frequencies so that they are closer to a harmonic series.

Equation 5.1 describes a mathematically ideal string for which the resonant frequencies really are a harmonic series. Real strings, like the ones we use on our guitars, don't quite make a harmonic series, but are usually close unless the player greatly shortens the effective length of a string by fretting it high on the neck. Harmonics of a real string are increasingly sharp, though the errors are usually small for the lower harmonics. This deviation from the ideal is somewhat predictably called inharmonicity.

Figure 5.5 shows the inharmonicity measured from a Taylor 710 for the two open E strings. The data is from a binaural head placed in a large hemi-anechoic chamber. The curves are power spectral density (PSD), which is a normalized relative of the FFT. The frequency resolution is high, so the PSD values, in dB/Hz, are small—the power has to be divided up into a large number of points. The horizontal axis is

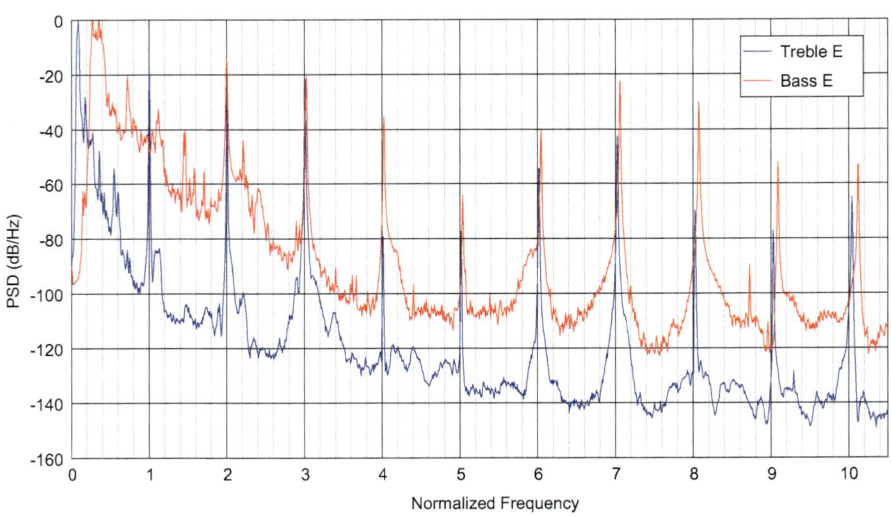

Fig. 5.5 Measured inharmonicity of a Taylor 710

normalized frequency, so that the fundamental is 1, the first harmonic is 2, and so on. The blue trace is the treble E string and the red trace is the bass E string.

Ideally, each peak should line up with an integer frequency. However, there is no way to fit a guitar with mathematically ideal strings. Alas, my local music store only sells real strings and they are all slightly inharmonic. The frequencies of the treble E string sharpen with each harmonic. The effect on the bass E string is more pronounced. Inharmonicity is a function of several parameters and one of them is string diameter, so heavier strings have more inharmonicity. The Bass E is wound, but its core has a larger diameter than the treble E.

These differences are not defects in either the guitar or the strings. The guitar is a particularly good one and it was fitted with high-quality strings. This is just physics. Inharmonicity is inversely related to length, so longer strings are closer to the harmonic ideal than short ones. This helps explain why concert pianos are so large. The longer strings sound better.

There are equations that more closely describe the behavior of real strings, but they are complicated and the ideal string equation is almost always used for calculating fret locations.

5.2 Vibration and Resonance

Engineers like to approximate complex objects with simpler ones that have about the same behavior, but are easier to describe mathematically. When talking about vibrations, they instinctively start with the simplest possible mechanical system, a weight on a spring with a damper. In textbooks, this is often called a spring-mass-damper system (Fig. 5.6). A damper is a component that absorbs energy, such as the

Fig. 5.6 A spring-mass-damper system

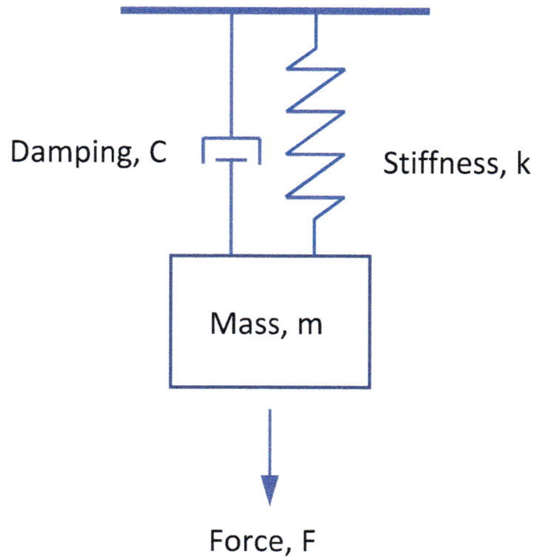

shock absorber on a car. In a guitar, damping comes from either the materials or energy lost to radiated sound.

With enough force applied to the mass, this simple mechanical system can be made to oscillate at any frequency. However, it is easiest to excite at its resonant frequency and will vibrate at this frequency if displaced and then released—akin to plucking. This frequency, in Hz, is

$$f_{\text{res}} = \frac{1}{2\pi}\sqrt{\frac{K}{M}}$$

Air is compressible and has mass, so an enclosed air mass can have a resonant frequency in the same way that a spring-mass-damper does. The simplest example is a Helmholtz resonator. This is a mathematical ideal with a resonant frequency that is easy to calculate. It consists of a volume of air enclosed by hard, rigid walls, and an open neck, as shown in Fig. 5.7.

The resonant frequency, in Hz, of a Helmholtz resonator is

$$f_H = \frac{c}{2\pi}\sqrt{\frac{A}{VL}}$$

where c is the speed of sound. At standard conditions—sea level at 20 ° C (68 °F)—the speed of sound is 343 m/s or 1125 ft./s. A guitar body isn't exactly a Helmholtz resonator, but is close enough that the idea of Helmholtz resonator is helpful in understanding how acoustic guitars work.

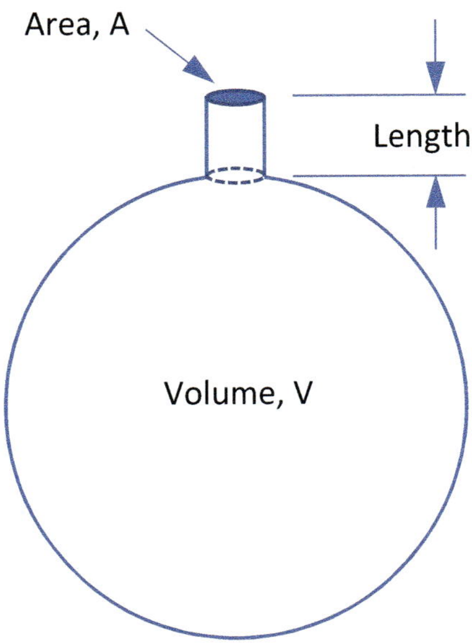

Fig. 5.7 A Helmholtz resonator

5.3 Resonant Frequencies and Mode Shapes

The literature on guitars is full FRF plots like the one in Fig. 5.3, and accompanying conversations about things like monopole and dipole modes. For this to make any sense, we need to make sure we are conversant in the idea of resonant frequencies and mode shapes. Physicists and engineers sometimes call these eigenvalues and eigenvectors, after the mathematical functions used to calculate them.

We've already looked at resonant frequencies. However, structures don't just like to vibrate at specific frequencies; they also like to vibrate in particular shapes at those frequencies. These are called mode shapes or, by the more nerdly among us, eigenvectors.

The discussion can get a little involved, so it's easiest to start with strings. Ideal stretched strings make harmonic frequencies and their corresponding mode shapes are made of fractions of sine waves. Say an A string is tuned to 110 Hz. If the string was excited at exactly 110 Hz (perhaps with an electromagnetic coil) and illuminated with a strobe light to "freeze" its motion, the shape would be half a sine wave (the blue curve in Fig. 5.8). Doing the same thing at 220 Hz, would give a full sine wave (the red curve in Fig. 5.8). The next resonant frequency gives one and a half sine waves and so on. The displacement of guitar strings is small, seldom more than 1 or 2 mm, so the amplitudes are exaggerated here for clarity.

Node points are places where the displacement of the string is zero. At the first resonant frequency of 110 Hz, there is no node point other than the ends of the string. For the second resonant frequency at 220 Hz, there is node point right in the middle of the string. This corresponds to the 12th fret. Since the string doesn't move at node points, the second mode can't be excited at the middle of the string.

As an experiment, pluck the open A string of an acoustic guitar where you normally would and then pluck it again at the 12th fret. The sound is clearly different and one reason is that all even numbered frequencies have node points there. The sound when plucked at the 12th fret is missing components $F_2 = 220$ Hz, $F_4 = 440$ Hz, $F_6 = 660$ Hz, and so on. Conversely, plucking that same string near the bridge produces a sound with lots of treble. This is because none of the lower frequencies have node points there, so more frequencies are excited.

You've almost certainly listened to demonstrations of node points on strings since some guitarists like to play harmonics. It's not too hard to place a finger lightly

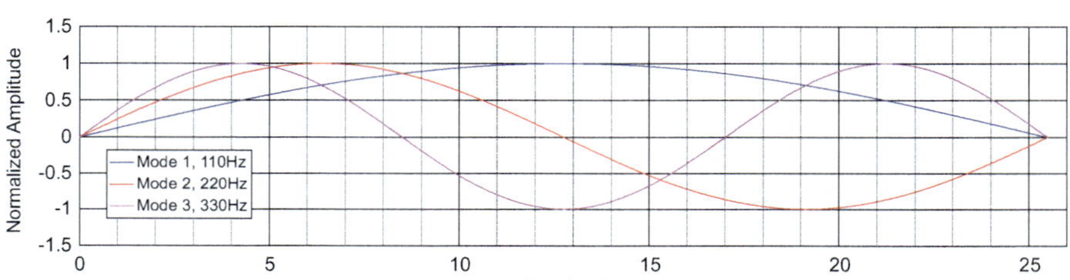

Fig. 5.8 First three modes of an ideal string, $L = 25.5$ in. = 648 mm

on a string at the 12th fret and pluck it while simultaneously releasing with your left hand. This makes a light, almost ethereal sound that skilled guitarists sometimes weave into their music. The finger at the 12th fret is at the node line of the second mode, so the player is exciting the second mode (and some higher ones), but not the fundamental. Additionally, players sometimes tune their guitars using harmonics.

Measuring mode shapes is more involved than measuring frequencies. Probably the simplest method of seeing the mode shape of a guitar top is to use Chladni patterns, made by sprinkling some kind of dust on a plate and exciting it at a resonant frequency. They are named after Ernst Chladni, a German musician and physicist who first described them in 1787, and are useful because they freeze mode shapes so they can be studied.

To make a Chladni pattern, simply sprinkle some light powder on a plate and excite it at one of its resonant frequencies. The powder gathers at the node lines where the plate isn't moving, "freezing" the mode shape. In the days before electronics and electromagnetic coils, people often bowed the edge of a plate with a violin bow. This required the plate to be firmly supported, often with a bolt through the center. Thus, the center of the plate was fixed and couldn't move—a node point. It takes some practice to apply the bow so that it excites the desired resonant frequency, but works surprisingly well when the test is set up carefully.

Figure 5.9 shows one of the mode shapes of a square plate along with the calculated mode shape. People wanting to take a deeper dive into plate vibrations can download Vibration of Plates by Art Liessa (NASA SP-160). Now 50 years old, it presents analytical solutions to entire families of vibrating plates and is widely referenced in the plate vibration world. A page turner, though, it is not.

On the right side of this figure, the points that aren't moving—the node lines—are in light green. The dark blue and dark red areas are the anti-nodes, the points where motion is a maximum. It's easy to excite motion at this frequency at antinodes, but impossible to do so at the node lines. Thus, the modes that can be observed depend on where the excitation is applied.

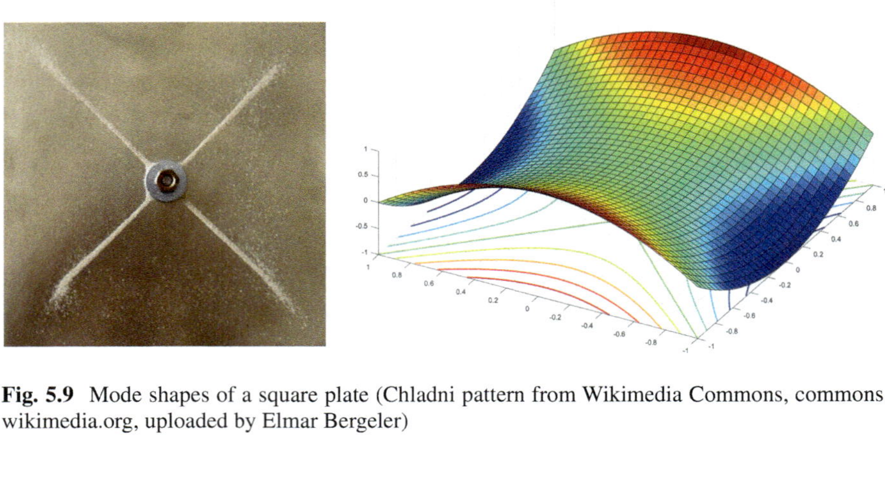

Fig. 5.9 Mode shapes of a square plate (Chladni pattern from Wikimedia Commons, commons.wikimedia.org, uploaded by Elmar Bergeler)

It's important to note that the displacement changes signs as you move across a node line. That is, positive displacement on one side of a node line means negative displacement on the other side. In Fig. 5.9, the maximum downward deflection is blue and the maximum upward deflection is red, with the green node line between them.

It's easiest to excite a mode at anti-node (this terminology is cumbersome, but we're stuck with it). In the case of the square plate, the violin bow could be applied at center of any of the four edges. Displacement there is maximum, so they are good places to induce motion. Conversely, trying to excite this mode near one of the corners won't work. The lines of powder show that the plate doesn't vibrate at those places and at this frequency. Bow all you want there and you might be able excite other modes, at other resonant frequencies, but you can't excite this one.

For top plates, back plates, or entire guitars, it is convenient to use a speaker for excitation. Typically, a computer or a piece of electronic equipment called function generator makes a pure sine wave at a desired frequency and the signal goes to an amplifier which, in turn, drives a speaker. Old references describe salt as the powder, but just about any light, visible granular material can work.

Figure 5.10 shows a nice Chladni pattern made by Trevor Gore. The resonant frequency is 289 Hz, as shown on the display and the strings are damped using energy absorbing foam ear plugs. Without this damping, the response of the instrument would be dominated by the strings and it would be very difficult to isolate the mode shapes of the top. Gore states that the testing material is a mix of old herbs found in his kitchen cabinet. He notes that fresh oregano also works well and smells good.

It's interesting that the mode shape is almost perfectly symmetric about the centerline of the instrument. Gore used a symmetric falcate bracing pattern on this instrument and that structural symmetry is apparent in the central node line.

It's important to note that a mode shape exists everywhere on the structure. People occasionally describe a mode shape as existing at some place on the guitar top, but not at others. This seems intuitive since a structure can't be excited at a node

Fig. 5.10 A Chladni pattern showing a mode shape on a classical guitar (image courtesy of Trevor Gore, goreguitars.com.au)

line and it seems like the mode must not exist there. It does, but has an amplitude of zero.

Chladni patterns are convenient since they are relatively easy to make, they "freeze" mode shapes in a very visible form, and they don't damage the instrument. In structural test labs, mode shapes are measured using more sophisticated methods, but they generate similar information. Figures 5.11, 5.12, and 5.13 show the first three modes of a Taylor 710 Dreadnaught guitar, measured using a device called a scanning laser Doppler vibrometer (SLDV). It measures time-dependent velocity of the surface at a grid of points and the software creates an animated mode shape from the results. The images here are frames from the respective videos.

The device on the right side of the lower bout is a small electromagnetic shaker that excited the structure by tapping it with a force sensor. The strings are heavily damped with a piece of soft foam just visible at the end of the fretboard, in the same manner as the ear plugs used by Gore in Fig. 5.10. The diagonal tubes behind the instrument hold blocks of soft foam against the back edges of the rim. The instrument was hung from a frame with rubber bands and these pads kept it from swinging without affecting the measured results.

The first mode has no internal node lines and second one has one node line (a curve actually) near the rim. Not until we get to the third resonant frequency is there an internal node line, roughly diagonal across the lower bout from lower left to upper right. This is an X-braced guitar, so the bracing pattern isn't symmetric about the center line. Thus, it's not too surprising that this mode shape isn't symmetric either. Note that the diagonal node line doesn't line up with either of the large X braces.

Fig. 5.11 Taylor 710 Top Mode 99 Hz

Fig. 5.12 Taylor 710 Top Mode 181 Hz

Fig. 5.13 Taylor 710 Top Mode 320 Hz

Descriptions of the physics of acoustic guitars often talk about monopole and dipole modes. Monopole modes are those with a single antinode and no internal node lines. Figures 5.11 and 5.12 show monopole modes. Figure 5.13 shows a dipole mode, as does the Chladni pattern in Fig. 5.10. Dipole modes have two adjacent areas moving out of phase with one another and separated by a node line. Out of phase means that one area is moving up while the other is moving down. In Fig. 5.13, the green and red areas are out of phase, moving in opposite directions. Dipole modes are often poor acoustic radiators since a portion of the soundboard making a positive pressure is right next to one making a negative pressure and the two tend to cancel one another out.

It's typical in full-sized acoustic guitars to have a first body resonance in the range of 95–105 Hz and the second body resonance in the range of 190–210 Hz. Many luthiers have target frequencies, though there is no universal agreement on what they should be. For example, a recent test on a pool of 12 classical guitars showed first resonant frequencies ranging from 92 Hz to 107 Hz and second resonance frequencies ranging from 191 Hz to 217 Hz. There was no clear relationship between resonant frequencies and subjective sound quality.

For most acoustic guitars, the first three modes follow this pattern, shown conceptually in Fig. 5.14. The first mode, often called the breathing mode, is one in which the top and back move opposite one another so that the interior volume changes as the body vibrates. This radiates low frequency sound from the soundhole. The second mode resembles the first, except the top and back move together. The top radiates sound, but there is much less sound produced directly from the soundhole. The third mode is typically the first one with a central node line.

The strings themselves don't make much sound because they have to so little surface area—they simply can't push much air around. To hear this effect, play an electric guitar without plugging it in. Rather, the sound of an acoustic guitar is produced by the vibrating soundboard and by direct radiation from the soundhole. The structure of the guitar slightly modifies the vibrations of the strings before radiating that motion into the air as sound waves. It's important to note that the effects of the neck and body are important, but small. If it weren't small, the instrument would tend to make sound at the body resonant frequencies rather than the string frequencies and it wouldn't be very useful for making music.

To illustrate, Fig. 5.15 shows a plot called a spectrogram of the sound from the same Taylor 710 shown earlier. The instrument was recorded in a high-quality

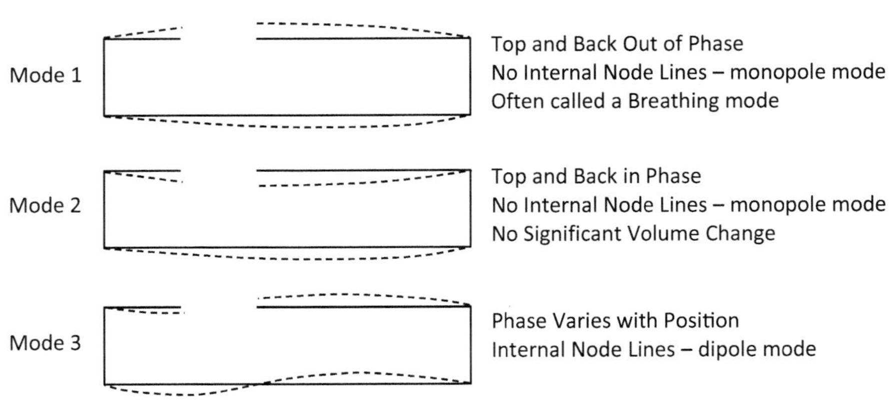

Fig. 5.14 First three modes of most acoustic guitars

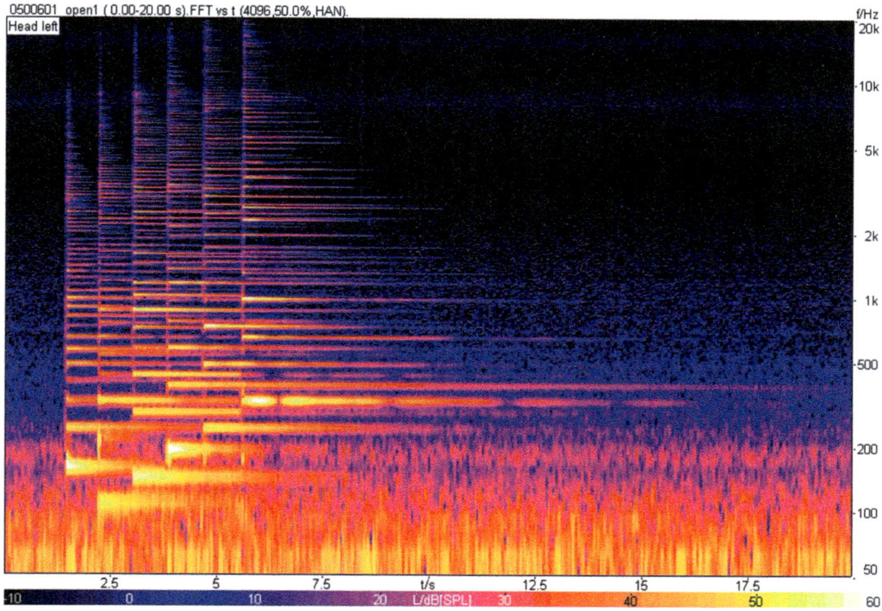

Fig. 5.15 Spectrogram of open strum of Taylor 710

hemi-anechoic chamber using a device called a binaural head that accurately simulates human hearing. The horizontal axis is time and the vertical axis is frequency, on a log scale. The colors indicate unweighted sound pressure level. The recording is of a slow open strum and it's easy to see the contributions of each of the six strings. This figure is dominated by the frequencies of the strings, whose response is tailored by the dynamics of the body.

The lesson here is that body resonance frequencies and their corresponding mode shapes are important. They strongly condition the sound quality of the instrument, but those effects are hard to see on a plot. The microphones used to make this plot are more sensitive and more uniform across the spectrum than any human's hearing and the listening environment was nearly ideal. Even so, the effect of the dynamics of the body are very hard to discern from the spectrogram.

5.4 Structural Coupling with Air

The dynamic behavior of acoustic guitars is highly dependent on the air enclosed by the body. It has mass and stiffness (since it is compressible) and can transmit waves, so it interacts with the vibrating structure around it. Luthiers instinctively know that the volume enclosed by the body and the area of the soundhole affect the sound and tailor those values in the pursuit of tonal targets.

As an experiment, tape a light piece of corrugated cardboard or foam core over the soundhole of an acoustic guitar and play. Simply removing the acoustic port without changing the internal volume makes a very noticeable tonal difference. As an example, the first two resonant frequencies of the Taylor 710 are 98.1 Hz and 181.9 Hz. Covering the soundhole eliminated the first resonant frequency and reduced the second frequency by 7.9% to 167.5 Hz. It still sounded rather like a guitar, but a different and less interesting one. I preferred the tone with the soundhole and apparently Taylor did as well.

While guitars without soundholes are rare, they do exist. Figure 5.16 shows an Alvarez Yairi DY88 dreadnought guitar without a soundhole. This instrument has a pickup and is intended for stage use; eliminating the soundhole reduces feedback in loud environments. Chris Smither notably played a blue Alvarez with an OM style body and no soundhole.

A cubic meter of air has a mass of about 1.2 kg (about 2.6 lb., more than the weight of a quart of milk). The air inside a full-sized guitar body has a mass of about 15gm (a little more than half an ounce), so it's small compared to the mass of the top and back, but it isn't zero. Since air is compressible, it has the equivalent of spring stiffness. Any system with mass and stiffness has a resonant frequency, so the air inside the guitar does as well. The lowest resonant frequency of the air inside a guitar body strongly conditions the low frequency sound of an acoustic guitar.

If you have blown over the open top of a bottle to make a whistle, you've found an acoustic resonance frequency of the air inside the bottle. You may even have

Fig. 5.16 Alvarez Yairi DY88 dreadnought guitar without soundhole (image courtesy of Mike and Mike's Guitar Bar, mmguitarbar.com)

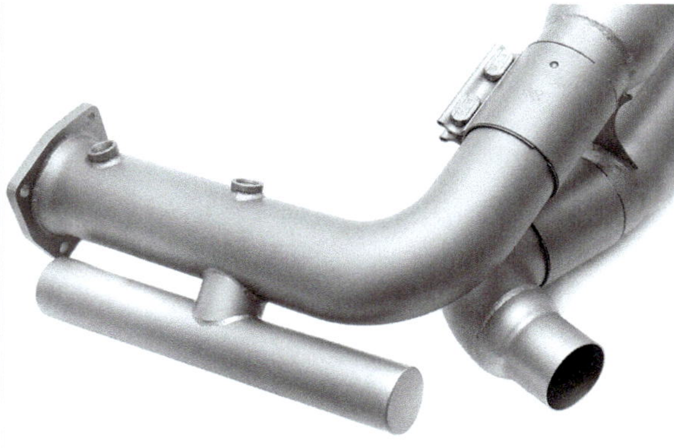

Fig. 5.17 A tuned resonator on the exhaust system of a sports car (image courtesy of Soul Performance, soulpp.com)

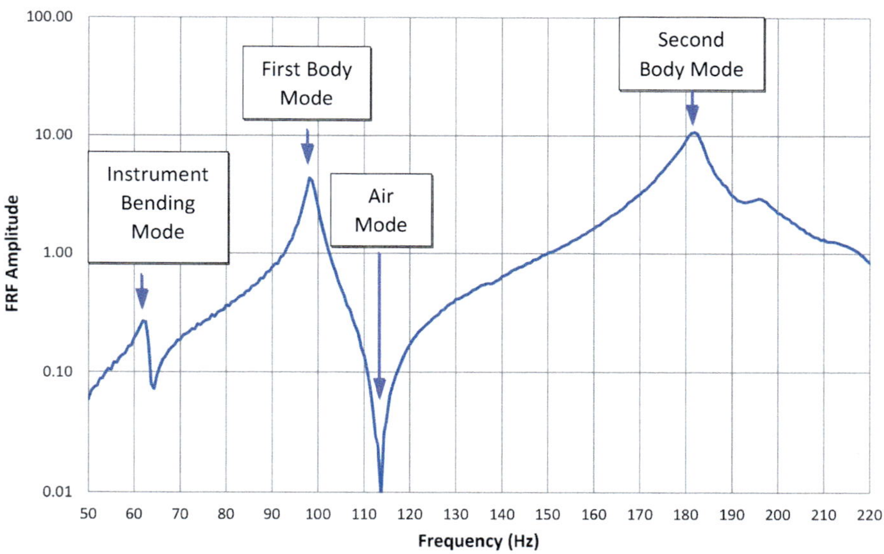

Fig. 5.18 Lower frequencies of Taylor 710

noticed that partially filling the bottle raises the frequency. If so, you showed that the geometry of a resonator determines its resonant frequency.

Acoustic resonators have an interesting property. Under some circumstances, they radiate sound at their resonant frequency, and in others, they absorb sound at that frequency. Buildings sometimes have Helmholtz resonators that act as absorbers at specific frequencies. Also, intake and exhaust systems in cars have tuned resonators to eliminate objectionable frequencies. Figure 5.17 shows a resonator on an aftermarket exhaust system for a Porsche 996, designed to attenuate an objectionable frequency at low RPM.

In most acoustic guitars, the first air resonance lies between the first and second body frequencies. Figure 5.18 shows only the lower frequencies of the Taylor 710

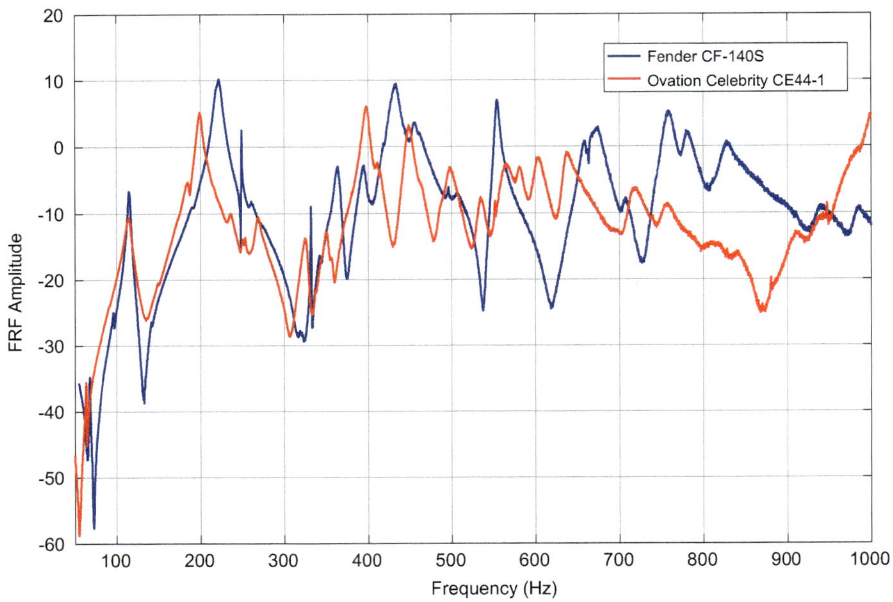

Fig. 5.19 Measured FRFs of two guitars

shown above. The "air mode" is at 114 Hz, between the first and second body modes. Note that it is a valley rather than a peak, as are the coupled structure-acoustic modes. The guitar actually absorbs energy at the air mode.

It's tempting to think that sound quality can be inferred from a frequency response function (FRF). They are certainly related and some luthiers have explored possible correlations. But, there is not yet any robust definition of measured guitar sound quality and no body of literature that reliably relates sound quality to specific features in FRFs. A convincing test would be a double blind one in which sound quality of a statistically meaningful pool of instruments was predicted based on carefully measured FRF plots. This would be followed by a subjective evaluation of the instruments. A statistically robust correlation would be a very important result.

For context, Fig. 5.19 shows the frequency response functions of two acoustic guitars. The first is a Fender CF-140S, an inexpensive and competent, but unremarkable folk guitar with a solid spruce top. The second is an Ovation Celebrity CE44-1, with a solid wood top and a deep bowl back made from Lyrachord, a proprietary composite material. They sound very different, so it's not surprising that the FRFs differ, but it's difficult to be more specific.

The curves in Fig. 5.19 are driving point FRFs measured at the bridges, just outside the saddles, as on the Ovation guitar shown in Fig. 5.20. The dot to the right of the micro accelerometer is the tapping point. The instrumented hammer is fitted with a soft tip to concentrate energy at the lower frequencies. The accelerometer is fixed to the bridge with a sticky wax. The blue tape just keeps the wax from being rubbed into the grain of the bridge.

Fig. 5.20 An Ovation Celebrity CE44-1 undergoing a tap test

Subjectively, the Ovation is more active at higher frequencies. It sounds brighter when fingerpicked and has a different voice than the Fender guitar. It's not surprising, then, there are many more peaks in the FRF for the Ovation in the range 300–700 Hz. The Ovation also has 15 small soundholes rather than the single large one on the Fender.

A simple way to explore the effect of the numerous soundholes is to cover some of them. Simple discrete models of acoustic guitar dynamics suggest that reducing the soundhole area should lower the first resonant frequency. To verify, 7 of the 15 soundholes were covered with corrugated cardboard, which is stiff and light (Fig. 5.21). The edges are sealed with heavy paper tape.

Figure 5.22 shows FRFs of the guitar with and without covered soundholes. The effect on the first frequency is significant and agrees with predictions from the discrete model, and there is a small shift in the second resonant frequency. However, agreement between 220 Hz and 900 Hz is quite good, in fact good enough that it would be difficult to tell which was which. The effect of the collection of small soundholes may be important to the sound of the guitar, but they don't affect the vibration of the structure at most lower modes. Thus, their affect can't be observed from this test.

Covering soundholes did not change the structure of the instrument. The cardboard is very light and taped to a part of the structure that vibrates very little. Rather,

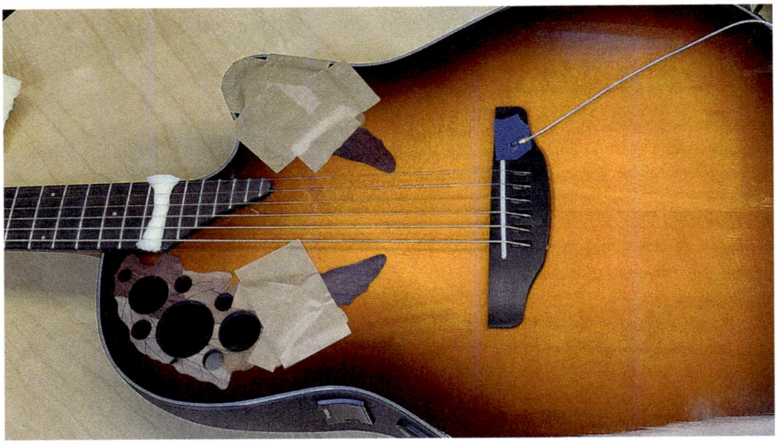

Fig. 5.21 Ovation Celebrity with partially covered soundholes

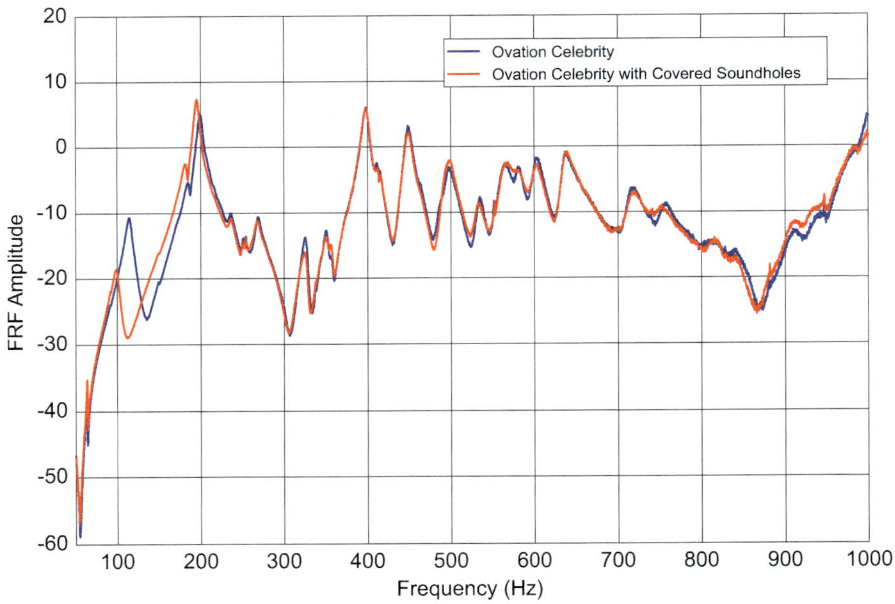

Fig. 5.22 Ovation Celebrity with and without soundhole covering

it changed the geometry of the soundholes, in turn changing the first acoustic resonance frequency (the Helmholtz frequency) of the body. Indeed, the Helmholtz frequency of the modified instrument just about matches the first body resonance frequency of the unmodified instrument.

In order to measure the acoustic resonance directly rather than infer it from the FRF, the top, back, and sides must be immobilized. One common way is to partially bury the body in a tray of sand, leaving only the soundhole exposed. The large additional mass and damping from the sand makes the top, back, and sides effectively rigid so that the only resonant frequencies are those of the internal air mass. There are numerous air resonance frequencies (theoretically, an infinite number), but the lowest one is of particular interest for guitar designers.

There is no easy way to calculate the resonant frequencies of the body structure or of the air. As one might guess, calculating them for the body coupled with enclosed air is even more difficult. In practice, they are almost always measured, though this could change as analysis tools improve. Until then, a helpful test is to make a fixture whose internal volume can be varied and on which the top can be changed quickly. This offers the chance to explore the coupled effects of changing body volume and soundhole diameter (Fig. 5.23). The fixture is made of MDF (Medium Density Fiberboard) and the body cavity is ringed with toggle clamps so that the top can be changed. The interior can be filled with MDF spacers to reduce the interior volume. In this image, the soundhole has not yet been cut.

To show the effect of interior volume and soundhole diameter, a series of sweeps was conducted by my student, Devon Pessler, who ran 10 different volumes and four different soundhole diameters, starting with a diameter of zero, as shown in the picture. There were four different tops, each with a heavy transverse bar between the upper and lower bouts to limit motion of the top bout. The bar is on the inside surface of the top in Fig. 5.23 and two of the MDF spacers are on the table next to the fixture. FRFs were recorded as body depth and soundhole size varied. The soundhole diameter varied from zero (no soundhole) to 3.75 in. (95.3 mm). Volume varied from 235.3 in.^{3} to 984.6 in.^{3} (3.86–16.13 L) (Fig. 5.24).

Increasing soundhole diameter tends to increase the first resonant frequency, roughly matching the behavior of an ideal Helmholtz resonator. Beyond that, though, there is no simple relationship between the body volume, soundhole diameter, and first resonant frequency.

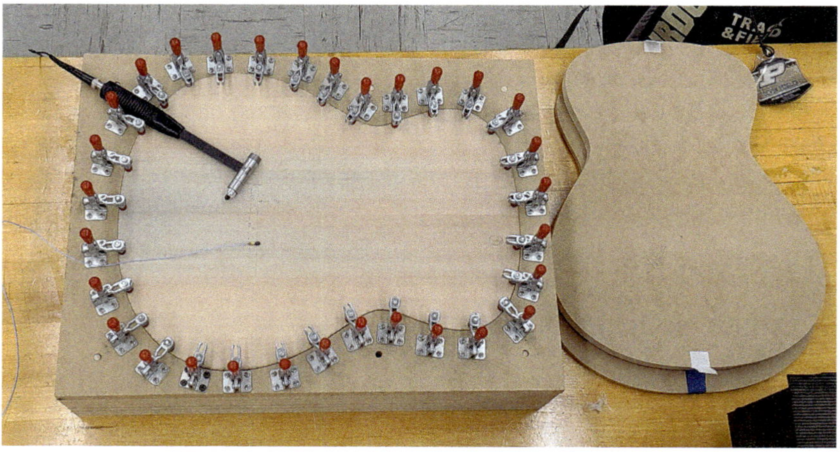

Fig. 5.23 Acoustic-structure test fixture

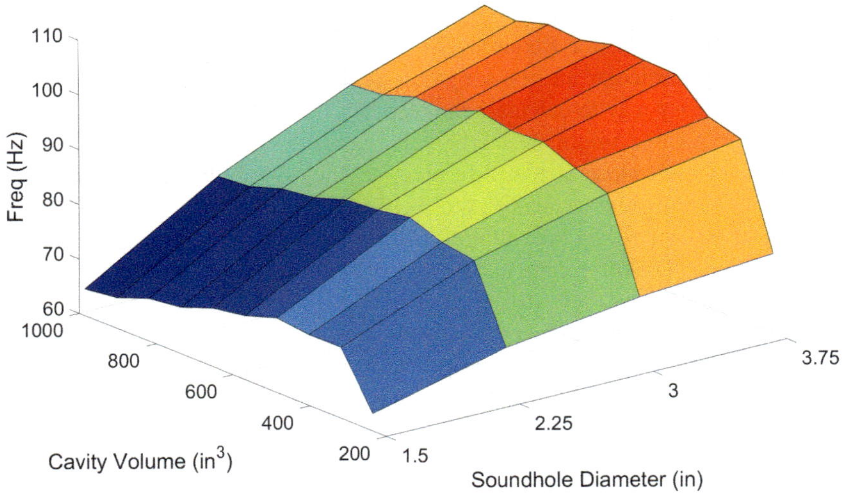

Fig. 5.24 First body resonance as volume and soundhole diameter vary

Luthiers have long known that larger bodies improve low frequency tone and it's no coincidence that upright basses have very large bodies and violins have small ones. However, the relationship between the first body frequency, the first air frequency, and the body volume in this test isn't enough to explain it. It's potentially important that the body volume in this test is only changed by varying body depth, with both length and width held constant. To make the point, Fig. 5.25 plots the first body frequency as a function of body volume, where body volume here is proportional to body depth.

This test highlights one of the ongoing challenges of using mathematical modeling for acoustic guitar design. The physics describing the coupled behavior of the structure and the enclosed air have been well understood for some time, but they can't be described accurately by a few simple equations. Rather, it requires sophisticated finite element models consisting of thousands of coupled equations. Researchers have shown that these models can accurately describe the behavior of guitars at low frequencies, but they are still expensive to create and their utility isn't yet worth the cost.

5.5 Mathematical Modeling

Many modern products are designed using software that predicts performance. Engineers then have a clear idea how their product will work before it is ever built. As an example, Fig. 5.26 shows the pressures calculated over the NASA X-57

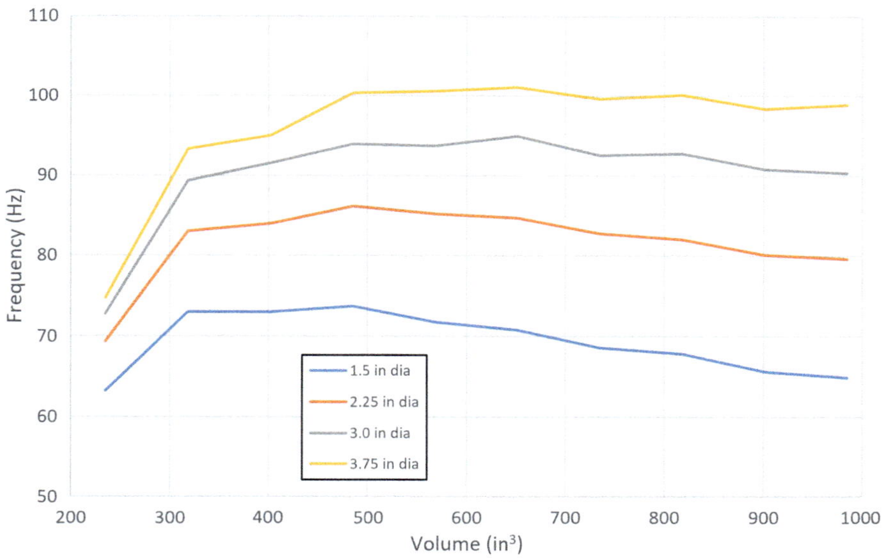

Fig. 5.25 First body frequency vs. body volume

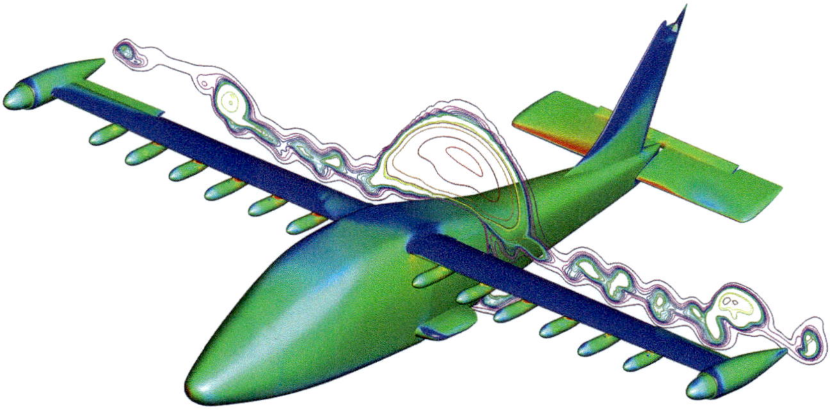

Fig. 5.26 Pressures calculated over the X-57 electric airplane (NASA, image is in the public domain)

electrically powered aircraft. These predictions matched test results well, something now so routine in the aerospace world that flight testing is often an uneventful verification of calculated predictions. In the aerospace world, people seldom build things they can't analyze.

Guitar design does not yet work this way, though the software does exist. The physics that make an acoustic guitar work are well understood and commercially available software can model those physics accurately. As an example, Fig. 5.27

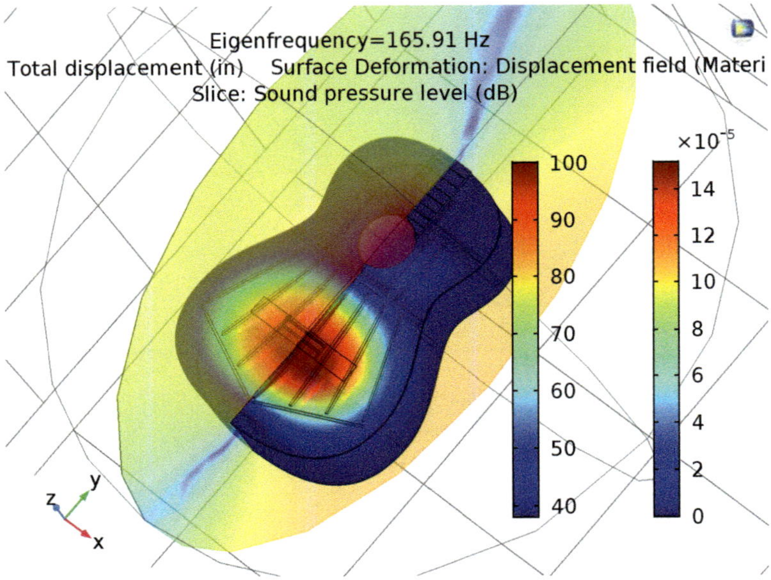

Fig. 5.27 Calculated mode shape and sound field around a classical guitar (image courtesy of Crisron Lucas and Franz de Leon, University of the Philippines)

shows the structural deflection and calculated sound field around a classical guitar, performed using a multi-physics package called COMSOL. In this case, multi-physics means that the motion of the body of the guitar, modeled by the equations describing the mechanics of solids, is coupled with the air inside and around the instrument, modeled by the equations describing how pressures propagate in air. This figure shows the second mode of the instrument.

Clearly, acoustic guitars can be analyzed at a sophisticated level, but almost no manufacturers or practicing luthiers do. There are two reasons. The first is that analyses such as these are still expensive. The software licenses are expensive because the software itself is so complicated, and it takes highly trained people a long time to make and refine these models. The first version of the model is often not accurate enough to be useful because it includes assumptions about material properties and other things. Rather, it must usually be tuned to match test results. Since manufacturers can make prototypes quickly, it is faster and cheaper to make test instruments as a way of refining a design.

The second reason that computational modeling is seldom used in design is that there is not yet a robust way of predicting sound quality from calculated results. The differences between an average guitar and a fine one are subtle. After the expense of making an accurate math model of the instrument, the problem remains of what to do with it. The designer would need to develop some way of relating calculated predictions to sound quality.

A less daunting step would be to use a math model to predict static stresses and deflections due to string loads. This would identify the most highly stressed parts of the instrument. Designers could then change the design to both reduce excessive deflections and to lower the maximum stresses. The result could be instruments that last longer and weigh less, as unneeded structure was eliminated.

Computational software will likely find its way into the guitar industry eventually. Fortunately, every generation of analysis software gets more powerful and easier to use. Large manufacturers will likely be the first to use engineering analysis as a routine part of the design process. The most dedicated individual luthiers may also explore uses for analytical software. It is easy to imagine that luthiers and manufacturers will develop empirical computational targets for their designs.

While sophisticated multi-physics models are still out of the reach of most luthiers, simpler models can be instructive. By far, the most familiar one is a simple discrete model proposed by Christisen and Vistisen in 1980. They treated the flexible top of the instrument as a mass supported by a spring representing the top stiffness. Acoustically, the body was modeled as a Helmholtz resonator. They presented a model with the top and the air mass in the soundhole as the two moving parts, called Degrees of Freedom (2—DOF model) as shown in Fig. 5.28. Another version of the model adds a flexible back, modeled in the same way as the top.

A representative FRF from this model is shown in Fig. 5.29. It clearly shows two peaks and the air mode, similar to the measured FRF shown in Fig. 5.18.

This discrete model is useful in showing how the flexible top couples with the enclosed air and doing it in a way that is very simple to calculate. The equations of motion are easy to program in even the most basic calculation software. There are numerous articles and books showing how this model and more sophisticated versions of it can accurately reproduce the first few peaks in a measured frequency response function.

The problem with this model is that it is very approximate and necessarily describes the function of acoustic guitars at only the most basic level. This was certainly the authors' intent and their work is rightly well-known. However, it's hard to apply when refining a guitar design. For that, more sophisticated models are needed.

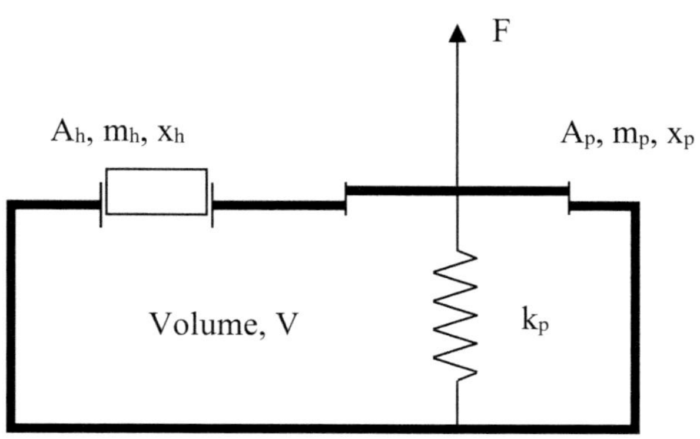

Fig. 5.28 The 2-DOF discrete model

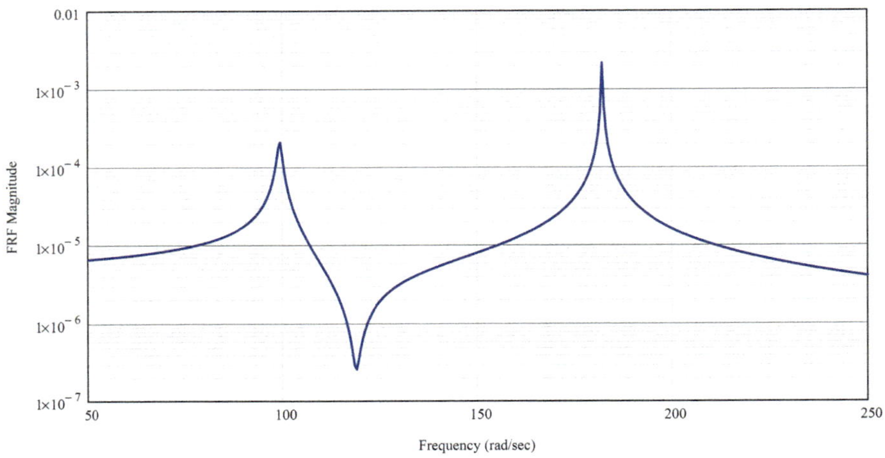

Fig. 5.29 FRF from the 2-DOF discrete model

5.6 The Mechanics of Glued Joints

Almost all guitars are made from wood and assembled with glue. No matter the type of glue, the mechanics of a wood-to-wood glue bond share a common model. Figure 5.30 shows a bondline model (adapted from a paper by Frihart and Beecher) that presents the important elements in a wood-wood bond. If any link in this chain fails, the joint can fail.

Frihart and Beecher offer some observations about failure in glued joints:

- Failure is seldom at the interface once a good bond is formed.
- Low density species are more likely to fail in the wood.
- Even strong wood species can fail in the wood if the cured adhesive is very strong.
- Adhesive interphase failure is common, though often attributed to interface failure because it can be difficult to detect a thin adhesive film on the failure surface.

A strong glue joint in wood relies on more than one type of adhesion and only one of them is the mechanical bond from glue integrating into the porous surface of the wood. The ideal joint fits tightly so that the resulting glue line is thin and even. The surface should be scraped or planed smooth and be free of dust, oils, or other contaminants.

People sometimes roughen the gluing surfaces, intending to give the glue more "grip." Data in the technical literature doesn't support the idea that a gluing face should be roughened or scored to promoted a stronger bond. Dr. Charles "Chuck" Frihart, from the US Forest Products Lab, observes:

> The purpose is to bring the actual substrate surfaces into close contact with the adhesive serving as a bridge between the surfaces. This means getting rid of the debris on the surface that can be a weak mechanical layer, such as damaged wood cells, or a chemically weak boundary with an oily wood like teak. Sanding by hand with a fine sandpaper is fine, but

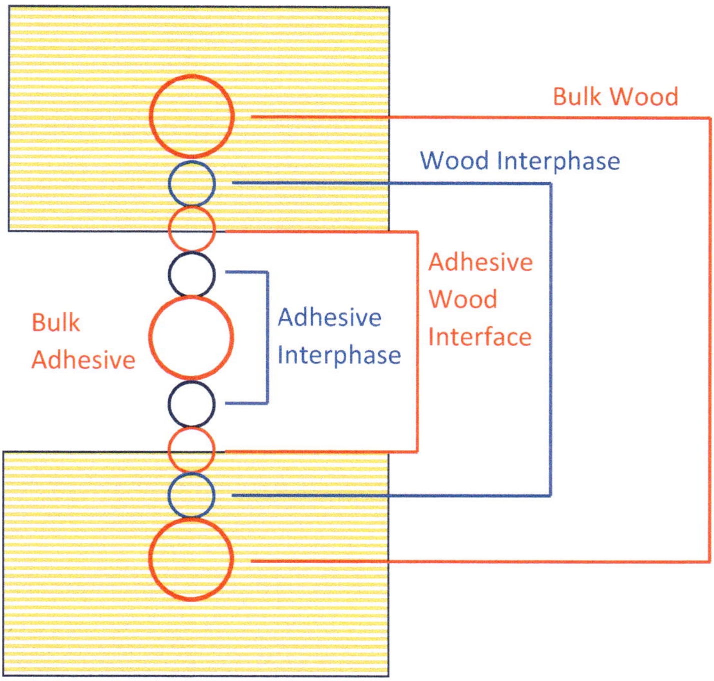

Fig. 5.30 Adhesive bondline model

rough sanding only weakens the surface layers compared to planing, gentle scraping, and light sanding with removal of the sawdust.

Some luthiers roughen backs of fretboards and bridges to reduce slippage when clamping. While this may indeed keep parts from sliding when clamped, it likely reduces the strength of the resulting joints. An alternate approach is to use alignment pins to hold components in place while the glue dries.

It is also critical that both surfaces are completely wetted so that there are no bare spots. This is complicated by the fact that wood absorbs water and can draw glue in, causing a glue-starved joint where too little glue remains to form a strong bond. In practice, many luthiers prefer to see a little glue squeezed out all along the edges of a glue joint. No squeeze out suggests that there may be a glue-starved area. Frihart cautions that defining a suitable glueline thickness is difficult because it depends on so many variables affecting adhesive-wood interaction. A very complete description of the physics and chemistry of glue joints is by Hunt, Frihart, Dunky, and Ruhoma (see the bibliography at the end of the chapter).

5.6.1 Creep

Creep in glue joints is the subject of a continuing discussion among luthiers. It's important to understand the engineering description of creep. Creep is easiest to describe using one of the simplest possible structures. Imagine a thin rod, clamped at one end and placed in tension by hanging a weight from it, as shown in Fig. 5.31. The initial length of the rod is L and its change in length is ΔL. Engineers define strain, ϵ, as ΔL/L, the change in length divided by the original length. ϵ is the lower-case Greek letter epsilon and is universally used in engineering books to designate strain.

Most metals don't creep at room temperature, so an aluminum rod placed under load that is not high enough to permanently deform it will just elongate and then stay in that state until unloaded. When the weight is removed, it returns to its original length. Engineers call this elastic behavior. When a load is applied to our aluminum rod, it behaves elastically, deforming to its new length and then stopping. However, some materials creep under load. This means that, even under load too small to cause immediate permanent deformation, they keep stretching over time.

Figure 5.32 shows how the rod would deform over time if it were made of a material that creeps.

The initial loading stretches the rod, but it continues stretching as time passes. Creep times are often measured in hours or days, and sometimes even in years. Since the working life of a guitar is usually measured in decades, creep can be a problem.

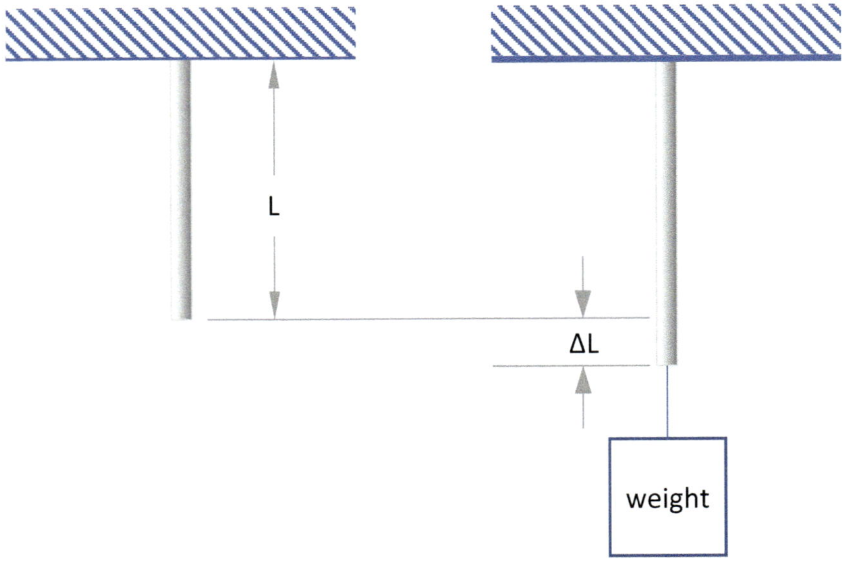

Fig. 5.31 A rod loaded in tension

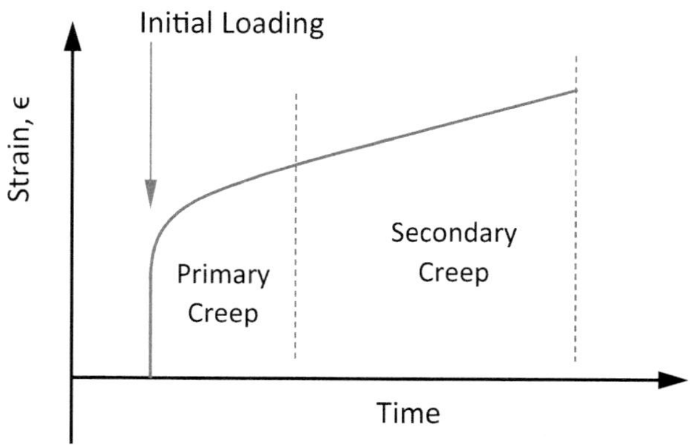

Fig. 5.32 Strain as a function of time in a material experiencing creep

Most metals don't creep unless they are heated far above room temperature, but wood and the polymers used in wood glue can. Since glue joints are generally in shear, there needs to be a definition for glue creep under shear loading. ASTM D7966/D7966M-16 defines creep in shear as shown in Fig. 5.33. Lines are scribed at the center of the glued faces and their movement after loading is recorded. Creep is recorded as the distance between the scribed lines. Several test conditions are defined in the standard. One, designated Test Condition A, is room temperature (20 ± 2 °C or 68 ± 3.6 °F) at 95% relative humidity. The shear stress at the glue face is 2.5 MPa or 360 psi, applied for 7 days—remember, creep happens slowly.

The standard was originally approved in 2014 and updated in 2016. Even so, there is little data in the literature that references it. Luthiers wishing to test to this standard should note that it defines a very specific pre-conditioning process for the specimens.

Luthiers should note is that shear creep is generally very small. The technical descriptions and informal discussions might suggest to a new luthier that a bridge could somehow slide across the top of a guitar due to creep in the glue joint. In practice, creep is usually measured in fractions of a millimeter or thousandths of an inch.

Dynamic loads are transmitted through glue joints, so it's important to consider their effect on vibration. If glue has an observable effect on the sound of a guitar, the most likely mechanism is increased damping.

5.6.2 *Viscoelasticity*

Some materials, including many glues, have a property called viscoelasticity. These materials lose energy to damping as they are loaded and unloaded. When force is plotted against displacement, the result is a loop, called a hysteresis loop. This loop shows how viscoelastic materials dissipate energy as they are loaded and unloaded.

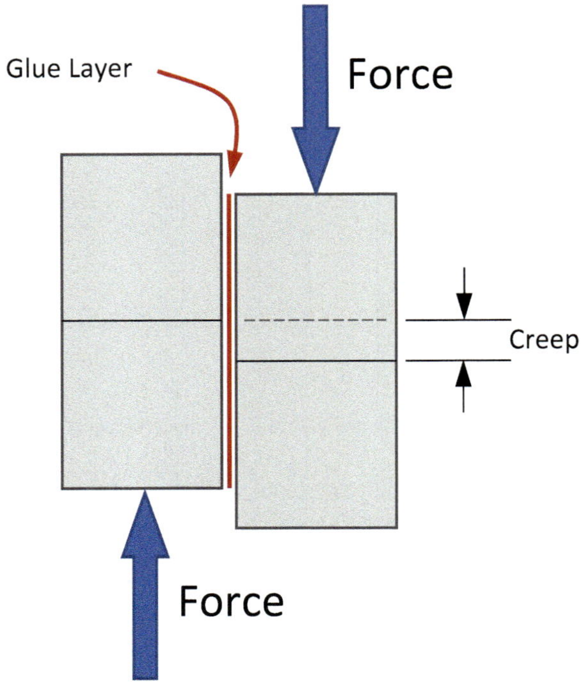

Fig. 5.33 Definition of creep in static shear (ASTM D7966/ D7966M-16)

The geometry of the loop depends on the material and how fast it is loaded. Figure 5.34 shows such a hysteresis loop. The area enclosed by the loop is related to how much energy is lost as a structure is loaded and unloaded as when it vibrates. Wood itself can be slightly viscoelastic.

Viscoelasticity greatly complicates structural analysis and mathematical models of structures generally assume the materials are elastic, so the hysteresis loop becomes a straight line whose slope is the elastic modulus.

One practical use of viscoeslaticity is constrained layer dampers. This is just a layer of a viscoelastic material, like butyl rubber, topped with a sheet of metal, usually aluminum. The aluminum top sheet and the panel to which to damper is glued combine to place the viscoelastic damping layer in dynamic shear as the panel vibrates. The damping layer soaks up energy and can drastically decrease the motion of the panel.

Patches like this are often bonded to the inside of car bodies to reduce vibration. There's a very good chance that your car or one you've ridden in has constrained layer damping patches in the body somewhere. They appear as patches on the inside of the stamped metal body panels. Figure 5.35 shows where they might be placed in a typical vehicle and a constrained layer damper placed inside a door panel. Dampers like this are one of the reasons that modern cars are so quiet.

Figure 5.36 shows a constrained layer damper on the inside of a door panel.

Fig. 5.34 A hysteresis loop for a viscoelastic material

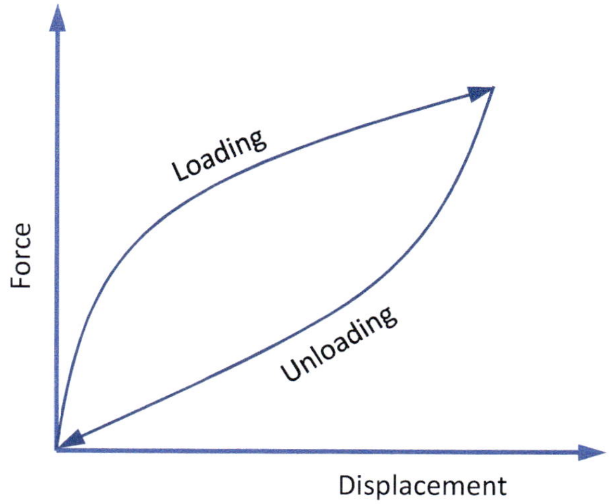

Fig. 5.35 Constrained layer damping patches in a vehicle body (image courtesy of Sika Automotive, automotive.sika.com)

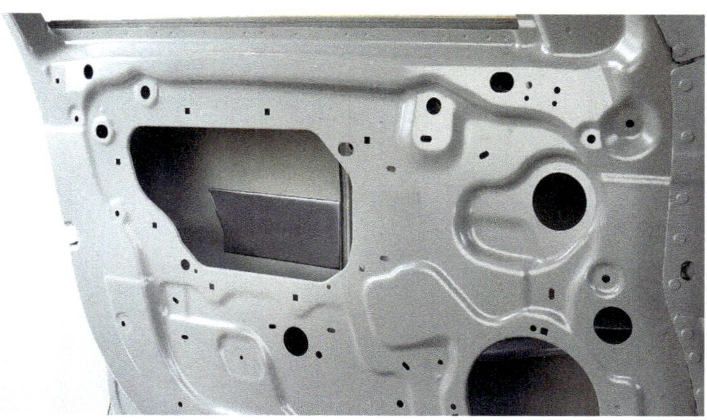

Fig. 5.36 A constrained layer damper on the inside of a car door panel (image courtesy of Sika Automotive, automotive.sika.com)

5.7 Mechanical Impedance

When two vibrating structures are mechanically connected, like strings attached to a soundboard, the ability of one to drive the other is defined by impedance. Mechanical impedance is mathematically similar to electrical impedance. For our purposes, it's best to skip the math and consider something less abstract.

As an example, imagine a circular plate free to rotate at the edges, but not to move vertically (engineers call this a pinned boundary condition). Since the first resonance frequency of the top plate of a full-sized guitar with the same boundary condition is in the neighborhood of 200 Hz, say that you'd like the first frequency of this circular plate to be 200 Hz. It's important to note that the top is pinned at the edges, but is not part of a guitar body, so there is no enclosed air volume acting on it. Figure 5.37 shows a disk with a slight upward deformation in the center (from an upward force). The lower bout of an OM sized guitar is 15 in. (381 mm) wide, so let's assume the circular plate has that as a diameter.

There is a simple equation for the first resonant frequency of a disk like this one, so it's easy to calculate results for disks made of different materials. The following plates all have a fundamental frequency of 200 Hz.

Material	Thickness (in)	Thickness (mm)	Weight (lb)	Weight (g)
Braced OM Top	n/a	n/a	0.386	175
Steel	0.266	5.74	11.33	5137
Aluminum	0.228	5.79	3.93	1783

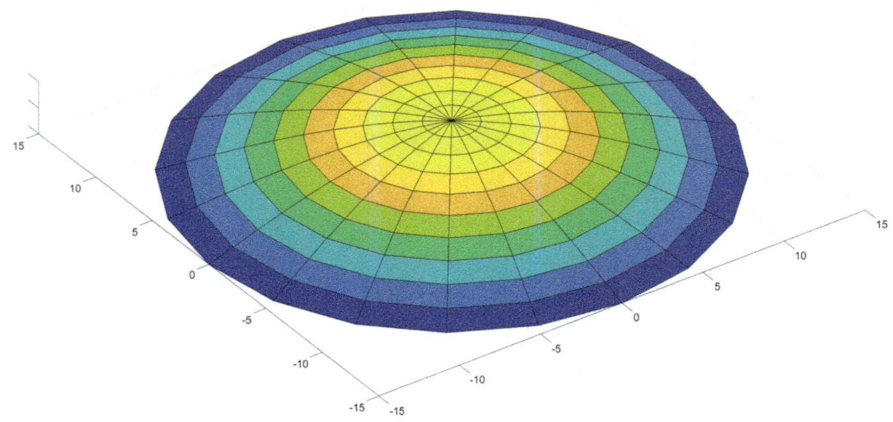

Fig. 5.37 Deformed disk

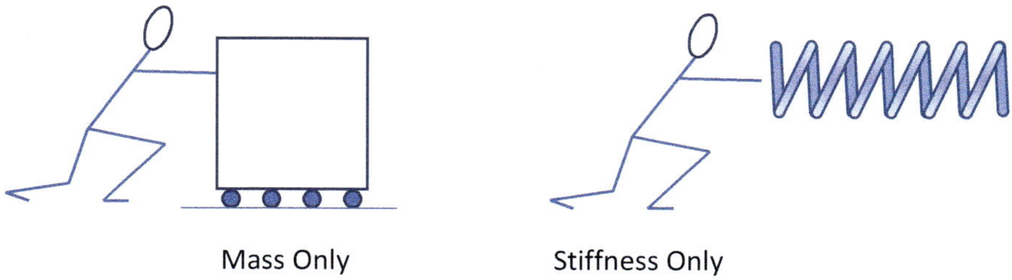

Fig. 5.38 Sources of impedance

They all have the same first resonant frequency, but vastly different mass and stiffness. Fixing six strings to the steel or aluminum disks would create very little sound because they are so stiff and massive that they wouldn't vibrate in response to the strings. Clearly, there is more to the story than just resonant frequency. That something is impedance.

Just as electronic impedance is the resistance to an oscillating voltage, mechanical impedance is the resistance to an oscillating force. In an acoustic guitar, the oscillating force comes from the vibrating strings. That causes the top to vibrate which in turn causes pressure waves in the air that we hear as the sound from the instrument.

Acoustic guitars parcel out the kinetic energy of the strings to the top and, through that, to the air. The details of how they do this determines the sound quality of the instrument. The mathematical details are quite complicated, but a few basic ideas are enough to make the point.

Mechanical impedance is a measure of how hard it is to make a structure oscillate and is related to the mass and stiffness. To start, imagine two structures with the same impedance, but very different characteristics (Fig. 5.38). The first is a large mass supported by wheels that are effectively frictionless. It has no stiffness since it will just keep rolling after it starts moving and won't return to its initial position. However, it is very heavy and hard to accelerate, so it has high impedance. The second structure is just a light, stiff spring. It has essentially no mass, but it is still very hard to move, so it too has high impedance. It's important to note that damping can contribute to mechanical impedance, but we're assuming that damping is low enough to be ignored.

Let's go back to our circular plates. They have the same frequency as a braced top, but are much heavier and much stiffer. Their mechanical impedance is so high that they would be useless as guitar soundboards.

Mechanical impedance is difficult to measure directly and it's not something a builder can track while working on an instrument. Fortunately, if the top has the desired fundamental frequency and is light enough, mechanical impedance doesn't have much choice but to come along for the ride. It's no coincidence that some of the best luthiers carefully record weights and resonant frequencies of their guitar tops. Individual builders often develop target weights and frequencies for their guitar tops.

I've skipped many of the details of mechanical impedance, such as the fact that it's a function of frequency. That means impedance is a curve, rather than a single number. Impedance curves also have peaks on them corresponding to frequencies at which the structure is particularly difficult to excite. For this discussion, the basic ideas are enough. The bibliography at the end of this chapter includes some references for people wishing to learn more.

Bibliography

Books

Bucur, Voichita, Handbook of Materials for String Musical Instruments, Springer, 2016, ISBN 978-3-319-32080-9

Fahy, F., Foundations of Engineering Acoustics, Academic Press, 2000

Fletcher, N.H. and Rossing, T.D., The Physics of Musical Instruments 2nd ed, Springer, 1998, ISBN 978-0387983745

French, R.M., Technology of the Guitar, Springer, 2012, ISBN 978-1461419204

Gore, T. and Gilet, G, Contemporary Acoustic Guitar Design and Build, 2nd ed, Published by Trevor Gore, n.d., ISBN Vol 1: 978-0-98711740-2-7, Vol 2: 978-0-9871174-3-4

Liessa, A., Vibration of Plates, NASA SP-160, n.d., retrieved from ntrs.nasa.gov

Articles

Christensen, O. and Vistisen, B.B., Simple Model for Lowe Frequency Guitar Function, Journal of the Acoustical Society of America, Vol. 68, 1980, p. 758, https://doi.org/10.1121/1.384814

Davis, Evan, On the Structural and Acoustic Design of Guitar Soundboards, Ph.D. Dissertation, University of Washington, 1990

Elkhateeb, A. and Eldakdoky, S., The Absorption Characteristics of Empty Pop Bottles, Acoustics Australia, May 2020.

Fliescher, H. and Zwicker, T., Investigating Dead Spots of Electric Guitars, Acustica – Acta Acustica, Vol. 85, 1999, pp. 128-135.

Firth, Ian M., Physics of the Guitar at the Helmholtz and First Top Plate Resonances, Journal of the Acoustical Society of America, Vol. 61, No. 2, Feb 1977, pp. 588-593.

French, R.M., A Pop Bottle as A Helmholtz Resonator, Experimental Techniques, May/June 2005, pp. 67-68.

French, M. and Bissinger, G., Testing of Acoustic Stringed Musical Instruments – An Introduction, Experimental Techniques, Jan/Feb 2001, pp. 40-43

French, R.M, Lydy, L. and Crawford, M., Comparing Subjective and Objective Data from a Pool of Classical Guitars. Savart Journal, North America, 1, Nov. 2019. Available at: <https://www.savartjournal.org/index.php/sj/article/view/28>.

Frihart, C.R. and Beecher, J.F., Factors That Lead to Failure with Wood Adhesive Bonds, Proceedings of 2016 World Conference on Timber Engineering, Vienna.

Hill, T.J.W., Richardson, B.E. and Richardson, S.J., Acoustical Parameters for the Characterisation of the Classical Guitar, Acta Acustica united with Acustica, Vol. 90, 2004, pp. 335-348.

Hunt, C.G, Frihart, C.R., Dunky, M. Rohumaa, A., Understanding Wood Bonds – Going Beyond What Meets the Eye: A Critical Review, Review of Adhesion and Adhesives, Vol.6, No.4, 2018, pp369-440, DOI: https://doi.org/10.7569/RAA.2018.097312

Jansson, E.V. "Fundamentals of the Guitar Tone"; Journal of Guitar Acoustics, No. 6, pp. 26–41, 1982.

Jansson, E., Acoustical properties of complex cavities. Prediction and measurements of resonance properties of violin-shaped and guitar-shaped cavities. Acoustica 37, pp. 211–221 (1977).

Ross, T.E. and Rossing, T.D. "Plate Vibrations and Resonances of Classical and Folk Guitars"; Journal of the Acoustical Society of America, 65, 72, 1979.

Rossing, T.D. "Physics of Guitars: An Introduction"; Journal of Guitar Acoustics, No. 4, 45–67, 1981

Torres, J.A. and Boullosa, R.R., Influence of the Bridge on the Vibrations of the Top Plate of a Classical Guitar, Applied Acoustics, Vol. 70, Issues 11-12, December 2009, pp 1371-1377, doi: https://doi.org/10.1016/j.apacoust.2009.07.002

Chapter 6
Detail Design

Preliminary design answers the big questions, like what kind of guitar it will be, what the basic structure will look like, and how big it will be. However, the details are what really turn a concept into a guitar. Indeed, the differences between a competent, but uninspiring instrument and a really good one are the result of getting the details right.

Many luthiers are familiar with the idea of a "10 foot guitar" or "3 meter guitar" if you prefer. From 10 ft. away, it looks like a guitar. Close up, though, it's not. There are guitars, including some attractive and interesting ones, that are so insufficient as musical instruments that they are effectively little more than decorations. This can be the price of getting the details wrong.

When asked for five important concerns for guitar designers, Kenny Hill, an accomplished builder of classical guitars, observed:

> #1 Neck angle and shape. It's quite different between steel string and nylon strings, but it is essential to getting the guitar to play well in the end.
>
> # 2 Accuracy of fretting and bridge position. Each of these things is about tuning and playability. Without these technical features in order, then
>
> #3 sound,
> #4 esthetic beauty or
> #5 craftsmanship
> don't mean a thing.

There are many small details that combine to create a successful instrument. For convenience, they are grouped here into a few broad categories.

The original version of this chapter was revised. The correction to this chapter is available at https://doi.org/10.1007/978-3-030-89381-1_8

R. M. French, *Acoustic Guitar Design*,
https://doi.org/10.1007/978-3-030-89381-1_6

6.1 Body

The guitar body has several jobs that sometimes conflict with one another. For example, it has to be strong enough to bear string loads and occasionally stand up to rough treatment while still being light enough and flexible enough to be the heart of a good instrument. The details of a well-designed guitar body are generally refined over several generations of prototypes. Indeed, even the most established designs are still evolving. For example, Martin still produces the classic OO-18 and they probably always will offer some version of it. The modern one looks very much like one made 75 years ago (Fig. 6.1). However, decades of small changes to both the instrument itself and the way it is manufactured mean that the modern one is more precisely made and probably exhibits less build variation than the original.

Before a design can be improved, it needs to be made repeatably and with little variation. Reducing build variation starts with clear, unambiguous geometric descriptions. One thing that's easy to overlook is the way guitar body shapes are designed and how those shapes are stored. Body shapes are designed using either traditional drafting methods or with CAD software. Body designs are mostly stored as either paper drawings or CAD files (CAD stands for Computer Aided Design). I use a concise mathematical definition of body shapes for my own instruments, but I am probably unique in that.

However body designs are developed and stored, they become hardware when used to make forms and templates or used to drive CNC machines. Any ambiguity in the way a design is defined or in the way it is stored risks creating variation in the final product.

Fig. 6.1 A 1946 Martin OO-18 (image courtesy of Mike and Mike's Guitar Bar, mmguitarbar.com)

This may seem picky, but it's important. A design that is poorly defined can't be accurately reproduced. Build quality and build variation are inversely related—the more of one there is, the less there is of the other. In practice, a luthier can only make cumulative improvements in a design if the effect of small, intentional design changes can be distinguished above routine variations due to the manufacturing process. This is at the heart of continuous improvement practiced across many manufacturing industries.

Most individual luthiers build guitar bodies with templates. Often, the body shape is simply drawn onto some durable material and cut out to form a template. A little care in making it can produce a nice template that can last for decades. It's common for body templates to include locations of braces, the bridge, soundhole, and other features. Typically, the template is used to make a form in which the body is built.

There is a long history of luthiers working this way and it can certainly work well. Figure 6.2 shows one of the body forms used by Antonio Stradivari during his long career. Violin makers use inside forms, so this is fundamentally different than the outside forms typically used by guitar makers. Stradivari experimented with

Fig. 6.2 A violin body mold by Antonio Stradivari (Image retrieved from Google Arts and Culture, artsandculture.google.com, image from Museo del Violino, museodelviolino.org)

different shapes, but many of his fine instruments were made using this form. This approach seems to have worked out alright for him.

Sometimes, a durable reference template is used to make working templates that might not be expected to have a long working life. Working templates may have extensive information written on them and may even be applicable to more than one design. Figure 6.3 shows a poster board template by Charles Fox that includes four body shapes along with soundhole and brace locations. This template is used across a family of designs because of careful design decisions made by Fox. The different but related body shapes share enough internal features that they are structurally almost identical and can use many of the same jigs and fixtures.

Figure 6.4 shows sides of an OM sized guitar in a conventional outside body form that was made on a CNC router. The top was glued on while the sides were still in the form so that the instrument retained the correct shape.

A more versatile approach is an adjustable form that can accommodate a range of different body geometries. Figure 6.5 shows an adjustable body form made by Charles Fox, intended for individual luthiers who wish to make different designs without having to make a collection of single-purpose body forms. Note the end blocks that match the body curvature. The neck block includes an aluminum channel. The neck will eventually bolt on to the body and the channel allows the pre-drilled neck block to be bolted firmly and precisely into the block into the fixture. Fox uses the same radii at the neck and tail for all his acoustic guitars, so only one set of blocks is needed, making this a particularly versatile fixture.

At this writing, accurate body templates are available from several good suppliers, so a luthier wishing to follow traditional designs can simply buy the needed

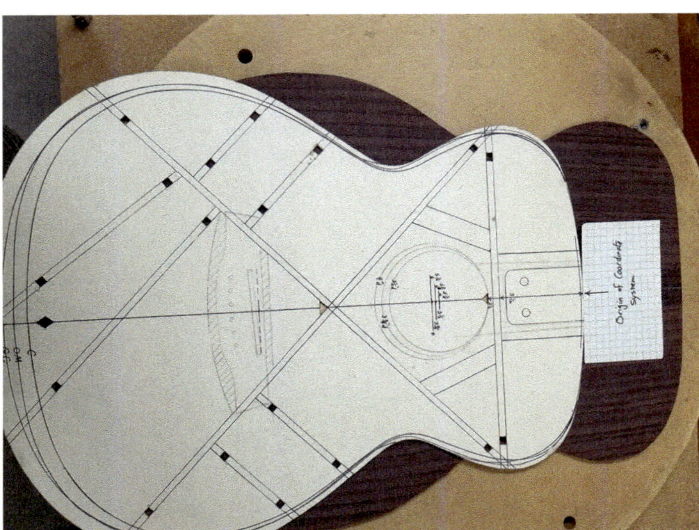

Fig. 6.3 A template by Charles Fox for multiple body shapes

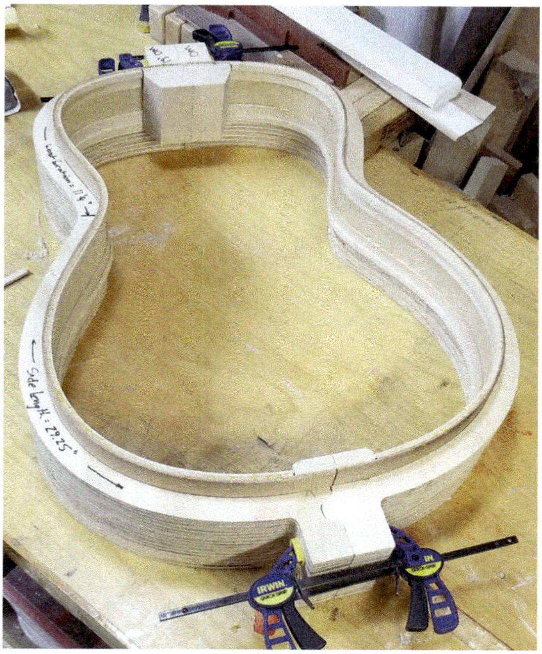

Fig. 6.4 Sides in a body form

Fig. 6.5 An adjustable body form

Fig. 6.6 A reference meter installed on the wall of the building (Wikimedia Commons, commons.wikimedia.org, uploaded by user Airair)

templates. This method is certainly workable. However, it doesn't support best practices for even small manufacturers.

Apart from the vulnerability of a design existing only in the artifact of the template, there is a loss of precision. If a luthier cuts out a master template and simply traces it whenever they need a new working template, error is introduced. If that second-generation template is traced, that adds another generation of error. A much more robust process is to define the shape mathematically rather than just as an artifact.

A useful analogy is from physics, where standards were defined by artifacts, but are now defined by fundamental and universal physical phenomena. For example, the meter was originally defined as the distance between two carefully machined lines on a bar made of platinum iridium alloy. This was an artifact and people wishing to measure distances precisely could only compare their own meter sticks to that bar, which was stored in Paris under the most careful conditions. The need for such standards is old. Figure 6.6 shows a reference meter set into the wall of the building in Paris in the 1790s. This public installation allowed people to compare their own meter sticks with a standard one.

In 1960, the definition of the meter was freed from a physical artifact and redefined as a certain number of wavelengths of a specific emission line from krypton-86. In 2019, it was changed again to be the distance light travels in exactly one second, making it dependent only on our ability to measure a second, since the speed of light (in a vacuum) is a constant everywhere. Luthiers certainly don't need to pursue this kind of precision, but the underlying philosophy applies. The way to move beyond templates as the definition of a design is to define the body of a guitar mathematically and the most popular way to do that is using CAD software.

CAD software is extremely common and some good packages are freely available. A body shape defined in CAD can be accurately reproduced whenever needed and losing or breaking a template becomes an inconvenience rather than a problem. Figure 6.7 shows the drawing of a template for a ¾ size guitar body. The drawing was used to cut a template from acrylic sheet using a CNC laser. Note this is version

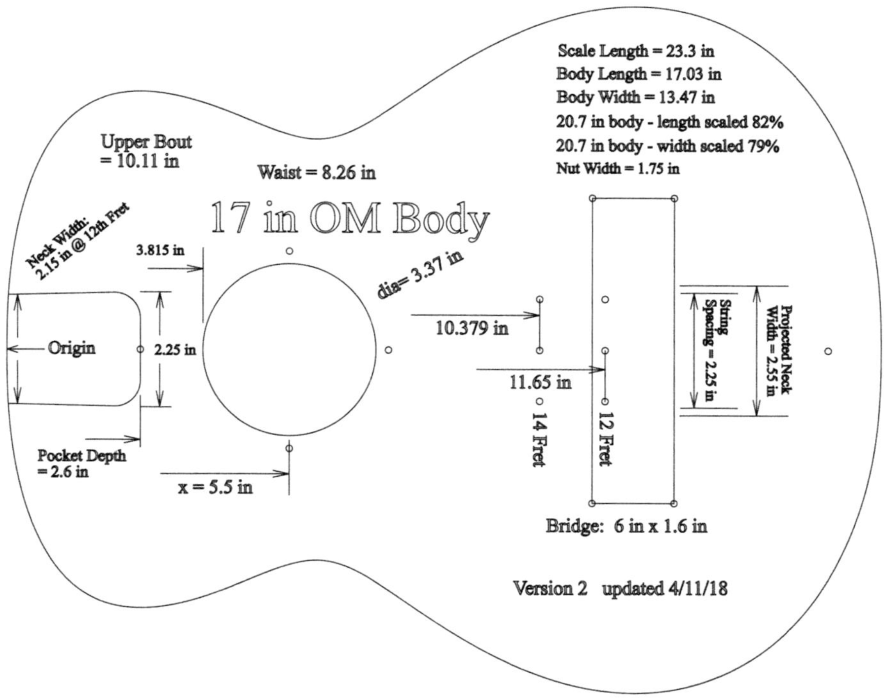

Fig. 6.7 CAD drawing of a small guitar body template

2 of the template, as noted directly on the template, along with some other useful information. Making it was simply a matter of opening the file for the original template, making the needed modifications, saving it as a new file, and cutting out a new template on a CNC laser.

Defining body shapes in CAD and then using a CNC machine to make the template ensures that every template is accurate and that all templates are identical to within the accuracy of the machine making them. It's also much easier to refine a design by working on a copy of the original file. Generations of refinements can be made with no loss of accuracy.

Within limits, a single design can be scaled to make a family of designs. As an experiment, I made a master body shape defined from instructions in Bob Benedetto's classic book on making archtop guitars. The body is about 20.7 in. (526 mm) long, so it's larger than most acoustic guitars. This is not a problem since the shape is easy to scale both the length and width to make a family of designs. The body form shown in Fig. 6.4 and the template in Fig. 6.7 both use this master shape. The template has the scale factors written right on it. Note that neck width doesn't scale with the body, so designers using this method must be careful in accommodating neck/body joint (Fig. 6.8, reapeated here for convenience from Fig. 4.44).

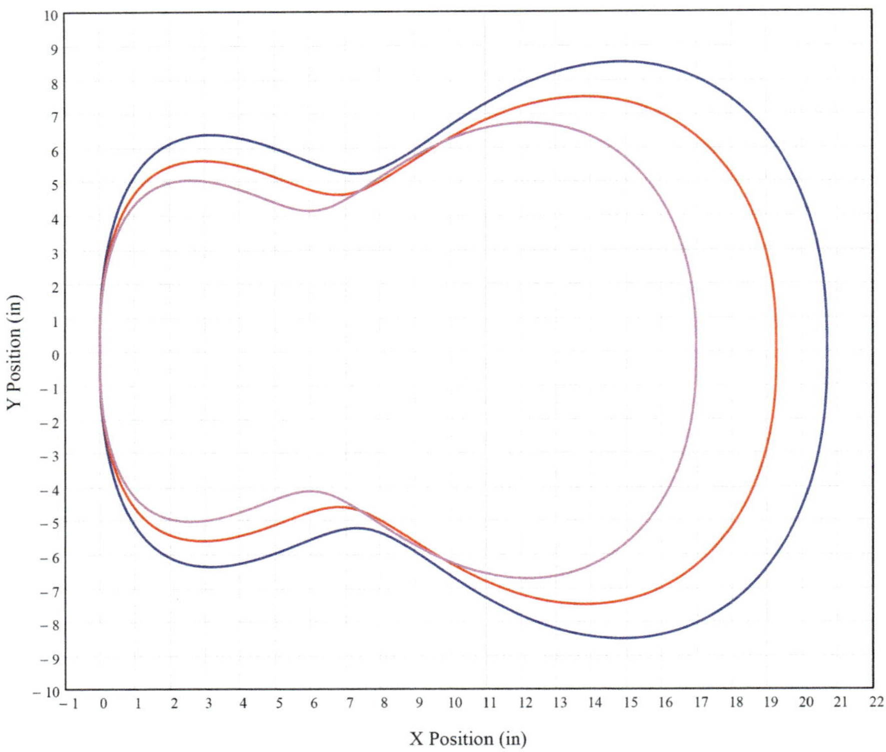

Fig. 6.8 A family of body shapes

If there is a problem with defining a body shape in CAD, it's that there is no closed form mathematical description of the body shape. It exists only as a list of coordinate points. The list of coordinates must be saved in a generic format and carefully backed up so that the curve can be reproduced whenever it's needed and isn't dependent on a specific piece of software.

A more fundamental way of describing a body shape is to write instructions for generating the shape. This is way, a shape can be precisely reproduced from a compact, unambiguous description without the need to push around a long list of coordinates. One traditional method is to define a body shape by a collection of circles joined by tangent lines. This way, a complete body shape can be defined by the center locations and radii of a few circles. Figure 6.9 shows a body defined this way from an article by Michael Darnton.

One way to improve repeatability is to account for differences in material properties. In an experiment here at Purdue, done with the help of my then student Eddy Efendy, tops for archtop guitars were milled on a CNC router using a fixture that drew a partial vacuum under the plates that slightly deformed them. Plates made of

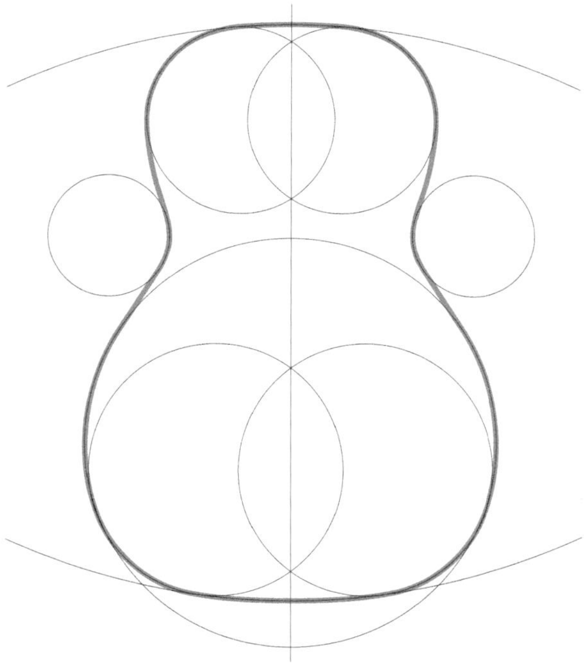

Fig. 6.9 A body defined using circles (image courtesy of Guild of American Luthiers, luth.org)

more flexible wood deformed more and, so, were slightly thicker. This increase in thickness increased stiffness more than mass and the result was that variation in the resonance frequencies of a group of test parts was lower for those tops machined with a vacuum than for those without it.

It's a fundamental rule of engineering that it's impossible to control things that aren't measured. Many of the best luthiers keep records of measurements from their instruments. These often include the weights of tops, often at different stages of building, and resonant frequencies. Keeping records like this doesn't need to be onerous and luthiers can develop a set of target values derived from particularly successful instruments. Having target values and refining instruments to meet them naturally reduces build variation.

Even basic measurements can be valuable. Luthiers sometimes keep small scales in their shops to record weight of tops at different stages of construction. Frequencies can be recorded easily using free apps on smart phones. Since frequency is related to mass and stiffness, it is helpful to have some top stiffness target. Again, this doesn't need to be complicated in order to be valuable. Trevor Gore suggests a target top stiffness that gives approximately a 2° rotation of the bridge under string tension, as shown in Fig. 6.10.

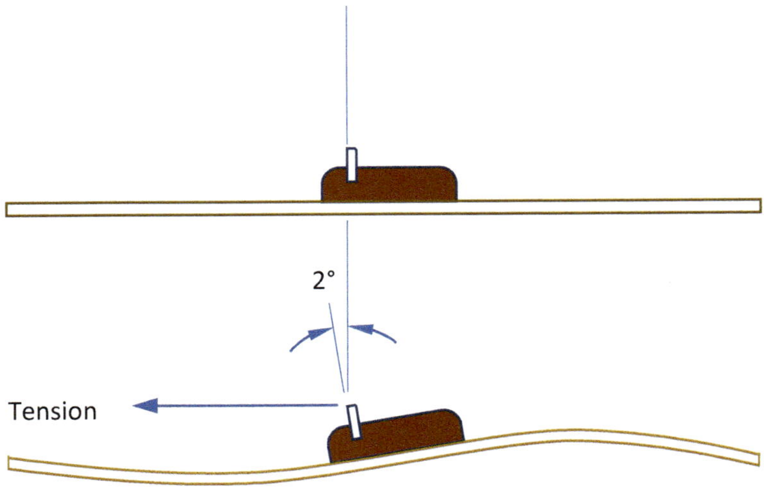

Fig. 6.10 Suggested bridge rotation under string tension

6.1.1 *Material Selection*

There is hardly a more energetic conversation in lutherie than that of selecting wood for a guitar. Selecting wood is the first step in making an instrument once the design is finished and the properties of that wood dictate many other decisions. It helps to have a clear starting point when selecting wood:

- Wood chosen for tops can have a significant effect on sound quality.
- Neck wood has a lesser effect on sound quality.
- Wood for sides and backs have minimal effect on sound quality.

A large majority of guitar tops are made from either spruce of a few difference species, cedar or mahogany. However, individual builders have probably used just about any wood imaginable. An extreme example is a student in the author's guitar manufacturing class who, ignoring every instruction from his professor, made an acoustic guitar with a top of walnut and purpleheart (Fig. 6.11). Both the purpleheart and walnut planks were flat sawn, with curved grain and shocking amounts of runout. Inexplicably, the guitar sounded good. This is why professors get gray hair.

6.1.1.1 Wood

There is slow progress being made on alternative materials for acoustic guitars, but for now they are almost all made of wood. Sides and backs are often laminated, even on mid-level production instruments, and some individual luthiers use laminated sides on their finest instrument. However, only the cheapest instruments use

Fig. 6.11 A guitar with a purpleheart and walnut top

laminated tops. The dynamic response of glued layers can so adversely affect the sound of the instrument that low cost and durability are the only attractions. For now, good sounding acoustic guitars almost all use solid wood tops.

If the dynamic properties of the top are so important in producing good tone, it is important to choose top wood carefully. One key feature is how the top was cut from the tree. Figure 6.12 shows the three directions of interest when cutting tops from a section of a tree: transverse, radial, and tangential. The longitudinal direction, not shown here, goes up the center of the tree.

The ideal wood for guitar top plates is quarter-sawn so that the growth rings are perpendicular to the faces. Quarter-sawn wood is less likely to twist or cup than flat sawn wood. Figure 6.13 shows a quarter-sawn plank in relation to the tree. Note that tangential shrinkage is about twice that of radial shrinkage. Thus, quarter sawn tops shrink less laterally, reducing stresses due to humidity change. Quarter sawn tops are also stiffer cross-grain.

Sections are cut from logs so that the transverse plane is perpendicular to the axis of the tree and a typical section is a little longer than needed for an acoustic guitar

Fig. 6.12 Radial, tangential, and transverse planes in a log (image courtesy of the DELTA CITES Wood ID project, delta-inkey.com)

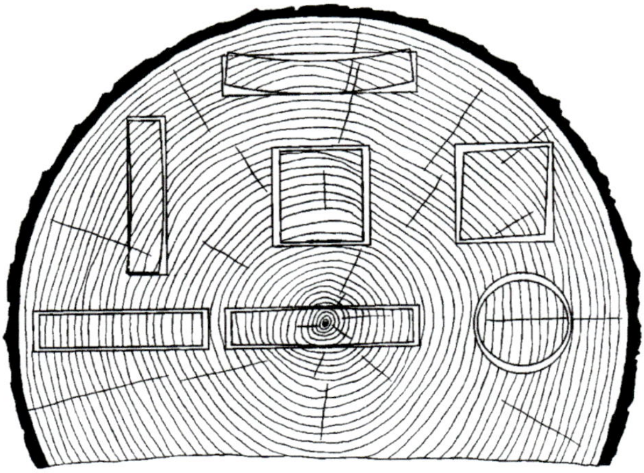

Fig. 6.13 Shrinkage and distortion due to drying (Handbook of Wood and Wood-Based Materials, Forest Products Lab)

top, with longer sections for larger guitars. Figure 6.14 shows some very large logs before being cut into sections. My former student, Erik Gallam provides scale.

Big suppliers must process large quantities of wood, usually, using proportionally large machinery. Smaller suppliers sometimes rely on more basic methods that allow careful attention to individual sections of logs. For these suppliers, the next step in processing is to split sections of the log into manageably small wedges. Figure 6.15 shows wedges at Alaska Specialty Woods being split using a traditional edged tool called a froe, and cylindrical hammer.

Fig. 6.14 Large spruce logs before being cut for guitar tops (image courtesy of Erik Gallam)

Fig. 6.15 Splitting sections of a large recovered Sitka spruce log (courtesy of Alaska Specialty Woods, alaskawoods.com)

Fig. 6.16 Sawing a wedge into top plates (image courtesy of Alaska Specialty Woods, alaska-woods.com)

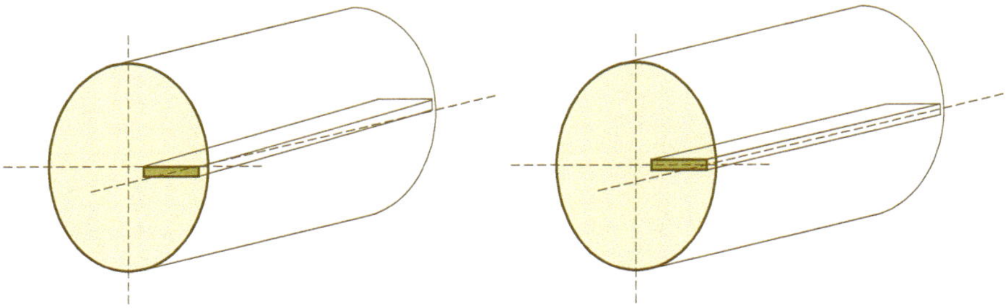

Fig. 6.17 Quarter-sawn boards with and without runout

Finally, the wedges are cut into thin top plates using a large bandsaw (Fig. 6.16). This process routinely produces nicely quarter-sawn top plates. This supplier specializes in salvaged and recovered woods, providing instrument-quality wood without having to fell living trees.

There are important differences in guitar top plates, even in perfectly quarter-sawn top plates. The radial plane should go through the centerline of the tree and be parallel to it. A top cut so that the radial plane is not parallel to the centerline of the tree has runout (Fig. 6.17). That means the wood fibers are not parallel to the face of the top plate, even though the growth rings are perpendicular to the faces. Splitting wedges along the grain and carefully resawing them reduces runout.

Grain runout appears on finished guitar tops as a difference in color across the center seam. Figure 6.18 shows an extreme case in which the grain of the top isn't straight and the runout changes directions between the bridge and the soundhole. While runout weakens the wood, it isn't necessarily a reason to reject a top and many good guitars exhibit some.

Fig. 6.18 Top of a classical guitar showing runout (Wikimedia Commons, commons.wikimedia.com)

Manufacturers necessarily buy their top wood in large lots graded by their suppliers, but individual luthiers often prefer to carefully select theirs. Some accomplished luthiers state that fine guitars can only be made from the highest quality wood, while other equally successful ones maintain that a skilled luthier can make a good instrument even from lower grade wood. Indeed, there is a long tradition of making guitars from gratuitously inferior wood, such as planks salvaged from shipping pallets. There is no common agreement on how to choose wood and research results do not always conform with popular practices.

While guitars have been made with what seems like every wood imaginable, there is rough agreement that tops should be made from spruce and cedar. Pacific Rim Tonewoods is a large supplier of guitar tops, producing about 300,000 tops yearly as of this writing. They state that 80% of their tops are Sitka spruce (Picea sitchensis).

Almost all acoustic guitars have either spruce or cedar tops. However, several manufacturers offer instruments with mahogany tops, like the Fender Hellcat shown in Fig. 6.19. This is a successful and modestly priced instrument whose tone reflects the choice of wood for the top. While a minority choice for guitar tops, mahogany has a long history, having been used in pre-WWII Martin guitars. Bob Taylor has even speculated that it would be possible to make a good guitar using nothing but mahogany.

When selecting tops, it's important to understand that wood is much more variable than materials produced by humans. For example, the mechanical properties of 4130 steel are well characterized and test samples display little variation, since the production process is carefully controlled and the formulation is clearly defined. Trees grow as they want to and there is large variation in mechanical properties within a species. There can be large variation even in a single tree. Thus, mechanical properties for wood exist only in ranges and tabulated values must be considered as only approximate.

Fig. 6.19 A Fender Hellcat guitar with a mahogany top (image courtesy of Fender, Fender.com)

Traditionally, luthiers have selected light, stiff wood with low internal damping for tops. In practice, this usually means wood with a tight, even grain and high surface hardness. Damping is often evaluated informally by tapping a plate with a fingertip while holding it lightly at a node point. Low internal damping allows the plate to ring longer.

Wood is a fiber composite and its mechanics are akin to those of a man-made unidirectional composite in which all the fibers are aligned in a single direction. As a result, there are separate values for lateral stiffness (across the grain) and longitudinal stiffness (along the grain). A common metric used to describe top wood is the ratio of longitudinal elastic modulus, E_L, to density ρ. The speed of sound in the longitudinal direction, c_L, is also the speed at which compressive strains propagate in that direction. It is defined as

$$c_L = \sqrt{\frac{E_L}{\rho}}$$

Some builders measure mechanical properties of tops and measuring longitudinal speed of sound can be an efficient way of quantifying an important property. Light wood with high stiffness along the grain has a higher speed of sound, c.

The speed of sound is proportional to the natural frequencies of a plate, so the larger c is, the higher the frequencies of the plate. Traditionally, luthiers have preferred wood with a high ratio of E_L/ρ. Indeed, this is one of the reasons often given for Sitka spruce being a preferred top wood.

A closely related metric is the sound radiation coefficient, R.

$$R = \sqrt{\frac{E}{\rho^3}} = \frac{1}{\rho}\sqrt{\frac{E}{\rho}} = \frac{c}{\rho}$$

Preferred top woods tend to be those with the highest sound radiation coefficient. Trevor Gore suggests that if a builder needs to trade stiffness against density for a soundboard, it is generally better to select the lower density top.

Fortunately, there are results in the literature that support these observations. An article by Merchel, Altinsoy, and Olson describes a large-scale test on a pool of spruce brace and top wood and its eventual use in five Taylor model 814ce acoustic guitars. Figure 6.20 is a plot from their article showing E_L vs. ρ for the tops. The tops for the five test instruments are shown by the circles along with their designators for each top. They selected brace wood using the same method and with the same distribution of E_L vs. ρ.

A carefully designed and executed listening test had the five instruments evaluated by both amateur and professional guitarists. A high-quality recording was made of a short chord progression played on each guitar. The group of test subjects listened to the recordings in a carefully controlled environment and rated the instruments. The results were analyzed statistically to ensure that listeners' ratings were valid.

Perhaps surprisingly, the guitarists preferred instrument Aa, with instrument Ab being only slightly less preferred. Instrument Cc was ranked lowest of the five. It may not be a coincidence that instrument Aa has the top with the highest R and instrument Cc has the top with the lowest R.

In practice, this means that good guitars may be made with lighter, more flexible wood. It could also lower the financial pressures on luthiers choosing to use wood of a lower grade, but high sound radiation coefficient. However, this would need to be acceptable to buyers conditioned to expect tops with closely spaced, even grain.

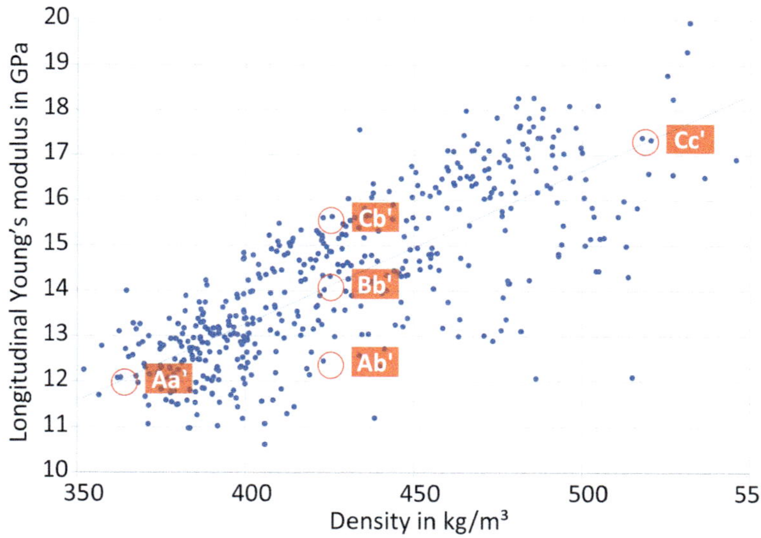

Fig. 6.20 E_L vs. ρ for a pool of spruce guitar top plates (image from article by Merchel, Altinsoy and Olson, courtesy of authors)

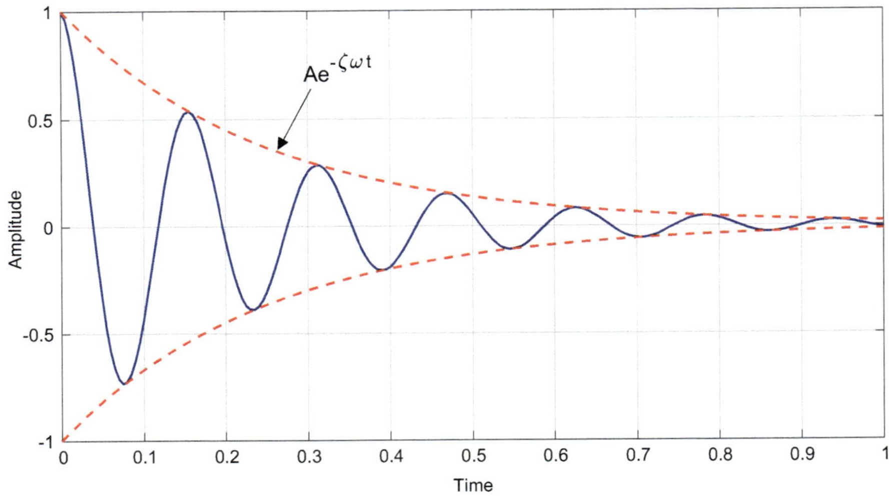

Fig. 6.21 Damped vibration showing damping coefficient, ζ

There is common agreement that wood with low internal damping is preferable. I know of no luthier who intentionally selects wood with a high damping value. Damping is usually defined in terms of either damping coefficient, ζ, or Q factor. ζ is the Greek letter, zeta. Figure 6.21 shows a damped sinusoid (solid blue line) along with its envelope (red dashed line). Note is that frequency, ω, is in rad/s, not Hz. The radian, abbreviated as rad, is the natural unit of angle in mathematics and there are 2π radians in a circle, so 2π radians = 360 °. Frequency in radians/s is just 2π times the frequency in Hz. 2π is about 6.2832. For example, the frequency of the bass E string is 82.407 Hz or 517.78 rad/s.

Luthiers often describe damping in terms of a Q factor that is defined as

$$2\zeta = \frac{1}{Q} \quad or \quad Q = \frac{1}{2\zeta}$$

Q and ζ are inverses, so high damping means high ζ and low Q. Conversely, low damping means low ζ and high Q. In Fig. 6.21, $\zeta = 0.1$, so $Q = 5$. Also, A = 1 and $\omega = 40$ rad/s.

Since any real vibrating structure has many resonant frequencies, it's very difficult to extract meaningful information from measured time data. It's much more common to look at vibration measurements in frequency domain, where it is easy to identify resonant frequencies. Figure 6.22 shows the previous function in frequency rather than time (blue), along with two other curves showing the effect of changing Q.

These factors are defined in texts on structural vibration and the underlying math can be a little involved. It's enough to know that high internal damping—high ζ and low Q—causes vibrations to decay quickly. Conversely, low internal damping—low

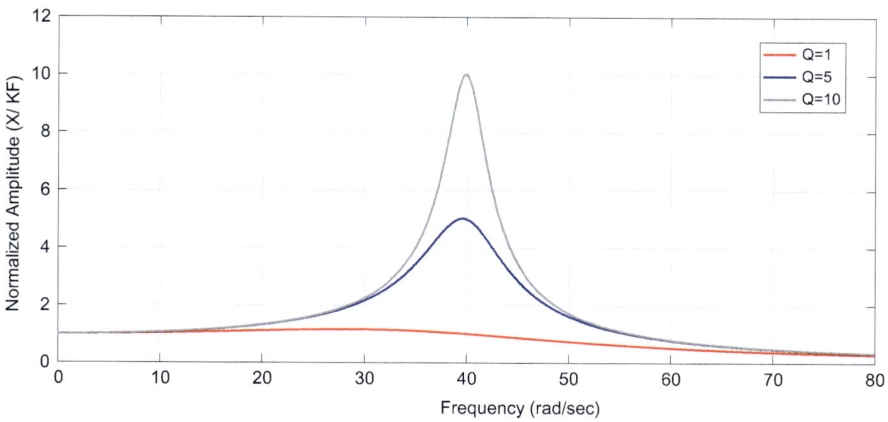

Fig. 6.22 Normalized frequency response function

ζ and high Q—allows vibrations to decay slowly. When someone taps a top plate and it "rings" for a long time, that's low damping.

Engineering texts on mechanical vibration tend to be dense and mathematically demanding. Rather, luthiers wishing to know more about the details of expressing vibrations mathematically should refer to Contemporary Acoustic Guitar Design and Build, Vol 1, by Gore and Gilet.

A final word on damping and stiffness of a guitar top comes from Andy Powers:

> The ratio of longitudinal to radial stiffness is a factor that translates to into how efficient a top is in it's unbraced, natural state. The longitudinal stiffness, with very few exceptions, will be higher than the radial stiffness. How much more determines how much the cross-grain stiffness must be altered via a bracing system to make a pleasant sound. If the longitudinal stiffness is low, in the range of the radial stiffness, the whole top has such high damping, the sound will be dull and uninteresting as in the case of a plywood top. If both the longitudinal stiffness and the radial stiffness are ultra high, the damping factor is so low at guitar note frequencies that musically unpleasant, inharmonic sounds are not removed and the guitar has a harsh, aggressive edge to the sound.

While top wood strongly affects tone, the wood chosen for sides and backs can be chosen more for stability and appearance. Wood with wavy grain is understandably popular, since it can be visually so striking. Figure 6.23 shows a Taylor 614ce WHB guitar with a highly figured maple back and sides. In guitars, this figuring is most often seen in maple and it has several common names. These include wavy, flame, curly, and tiger. Since violin backs are often made with highly figured maple, it is also sometimes called fiddleback maple.

There are several types of figuring in wood, including wavy or flame, quilt, and birdseye. These are due to genetic variations that cause the grain of the tree to not grow straight. Figure 6.24 shows a wood sample with a very strong wave. While this wood can be visually very appealing, it may reduce stiffness. There is little data on how figuring affects the tone of the instrument, particularly when used in sides and back.

Fig. 6.23 A Taylor Model 614ce WHB with figured maple back and sides (image courtesy of Taylor Guitars, taylorguitars.com)

Fig. 6.24 Highly figured wood (Wikimedia Commons, commons.wikimedia.com)

Figuring in top wood is less common and more subtle. Some spruce displays figuring that is usually called bearclaw. This is supposedly because people thought the markings were caused by bears scratching the trees. This is a great story and it might even be true. However, the figuring in bearclaw spruce is also due to distortions in otherwise straight grain. Figure 6.25 shows a guitar with a highly figured bearclaw spruce top. There is little data to suggest that bearclaw significantly affects the tone, though it is a topic of discussion among luthiers.

Traditionally, guitar sides were often made from Brazilian rosewood, but this wood is now endangered from overharvesting. It is now very expensive and subject to restrictions from the CITES treaty. Where rosewood is still used, it has been replaced by East Indian rosewood, a more sustainable alternative. Rosewood is easy to bend and resistant to cracking over time, so it's an attractive choice.

Fortunately, the structural demands on the sides and back are low and fine guitars are routinely made with a wide range of back and side woods. When sides do

Fig. 6.25 A classical guitar with a highly figured bearclaw spruce top (image courtesy of Alaska Specialty Woods, alaskawoods.com, instrument by Paul Beauchamp, beauchampclassicalguitars.co.uk)

Fig. 6.26 A cross-grain brace being installed

experience structural failure, it's often cracking along the grain. Sides also sometimes distort in areas where the curvature is low. Figure 6.26 shows a brace being glued onto the uncurved section of an East Indian rosewood side, in a class taught by Charles Fox. Note that your author did clean up all that excess glue after taking the picture.

Fig. 6.27 A laminated side with a solid lining

The stiffness of the sides can slightly affect tone and some accomplished luthiers build additional stiffness into their sides. One popular approach is to laminate the sides from two or more layers. Figure 6.27 shows a side with an inner layer of 1 mm, three ply aircraft birch plywood. The outer plies are vertical and perpendicular to the grain of the solid outer layer. The outer layer of solid wood is about 2 mm (0.079 in.) thick. This side also has a bent solid lining rather than slotted kerfing.

A different structural approach is to make the sides using sandwich construction, a feature often seen in aerospace structures that need to be both strong and light. Figure 6.28 shows a side made by Charles Fox that has a thin solid wood outer layer and a thin ply inner layer, separated by a layer of rigid foam. In addition to being light and rigid, it should be very stable over time.

6.1.1.2 Alternative Materials

A large majority of acoustic guitars are made of wood, either solid or laminated. The most common alternative materials are carbon fiber (graphite) and high-pressure laminate (HPL) made from layers of paper and phenolic resin. However, a few manufacturers are exploring other options. One of these is a linen-based composite. Figure 6.29 shows an instrument from Blackbird Guitars. The fibers are from

Fig. 6.28 A side using sandwich construction

Fig. 6.29 Blackbird El Capitan Guitar made from linen composite (image courtesy of Blackbird Guitars, blackbirdguitar.com)

Fig. 6.30 A molded headstock on a Blackbird El Capitan Guitar (image courtesy of Blackbird Guitars, blackbirdguitar.com)

the flax plant and have the same structural role as the carbon fibers in a graphite guitar. The resulting composite, called Ekoa®, is intended to be both sustainable and durable. It is used throughout the instrument.

Being able to mold components, whether made using linen or graphite fiber, offers design options that aren't available to makers of wood guitars. Blackbird guitars have a hollow neck and this model also has a molded headstock, as shown in Fig. 6.30. The designers have taken advantage of the unique properties of the material by creating structural elements that wouldn't be possible with wood.

6.1.1.3 Glue

All wood guitars are glued together and it's important that designers understand the properties of the glue they decide to use. Only a few types of glue are routinely used in guitars and they almost all are either animal glue, aliphatic resin (yellow wood glue, abbreviated as AR), or epoxy. AR glue is by far the most widely used since it is strong, easy to use, inexpensive, nontoxic, and has a long shelf life. Hot hide glue is still widely used in classical guitars and occasionally in steel string instruments. Epoxy is less common though plenty strong for general use in instruments and is sometimes used as a grain filler. Composite instruments are made from fibers (usually carbon fibers) impregnated with epoxy, often a thermoset that cures when

Fig. 6.31 Dry hide glue granules (Wikimedia Commons, commons.wikimedia.org, image uploaded by user LivingShadow)

heated. Cyanoacrylate (super glue) is used for smaller features and for repairs, but seldom for major structural elements.

The oldest adhesive used in guitars is hide glue, a member of a larger group of animal glues. These are made from collagens found in skin, bones, and other tissues. Hide glue is usually supplied as dried granules (Fig. 6.31) that are rehydrated and then heated to liquify it. It hardens as it cools and dries.

Several manufacturers offer hide glues that are liquid at room temperature because of the addition of urea or other chemicals. These have not been widely used by instrument makers, though they are sometimes used for repairs. At this writing, at least two suppliers offer a liquid fish glue used for repairs.

Animal glues have been used since ancient times. While they have been largely displaced by modern polymer glues, hide glue is still used in a few specialized applications. Musical instruments may be the most common, but it is also used by conservators. In guitars, hide glue is most often used in classical guitars.

Yellow wood glue (e.g., Titebond) is aliphatic resin (AR), related to polyvinyl acetate (PVA), the white craft glue that we all used in elementary school. Perhaps the most familiar brand of white PVA is Elmer's Glue-All.

PVA is reasonably strong, safe, and easy to use, but is not suited to instrument making. It remains soft when cured and creeps under load. Wood glue, dyed a light yellow to distinguish it from white PVA glue, retains the attractions of PVA while being more heat and water resistant. It also dries harder and is more resistant to creep.

There is debate about the extent to which glue creep causes problems in guitars. Seams, even when tight and thoroughly wetted, can move enough to create a slight discontinuity. This can crack brittle finishes. This is sometimes evident on laminated necks and headstocks, such as the one in Fig. 6.32. This differential expansion

Fig. 6.32 A headstock glued from several pieces of wood

or contraction may be due to changes in humidity affecting the various woods slightly differently. RM Mottola has noted a slight bump in joints of laminated necks made with AR wood glue, an effect not present in those made using epoxy. This may be due to AR glue wetting the wood at the seam.

Titebond brand wood glue has been widely used throughout the guitar industry for decades. Franklin, the manufacturer of Titebond, states that Titebond I, II, and III are all subject to creep, though Titebond I is most resistant. The increased water resistance of formulations II and III comes at the expense of decreased creep resistance. They offer formulations with an extended open time for complicated assemblies. One of these, Titebond I Extend, has the highest creep resistance of their AR wood glues. Hot hide glue is generally considered to resist creep better than some formulations of AR glue, though there is not a large body of data in the technical literature.

Glue is the subject of an ongoing discussion in the guitar world that often revolves around the relative merits of hide glue and AR. Fine guitars are routinely made with both types and your author is agnostic on the topic. The important elements of the conversation include the following:

AR Wood Glue

- Safe—approved for indirect contact with food, very low volatile organic compounds (VOCs).
- Shelf-stable—lasts for months in the bottle when stored properly.
- Strong—stronger than wood when used correctly.
- Convenient—simply applied from the bottle and allowed to dry.
- May creep over time—joints may slip, cracking finish.
- May add damping—little hard data in the literature.
- Doesn't stick to itself—separated joints must have glue removed before regluing.
- Joints and unintentional smears can be visible under finish.

Hot Hide Glue

- Safe—related to gelatin, non-toxic and no VOCs.
- Strong—stronger than wood when used properly.
- Sticks to itself—separated joints can be heated to reactivate the glue.
- Archival durability—proven to last hundreds of years in stringed instruments.
- Assumed to add little damping when used properly—little hard data in the literature.
- Resistant to creep—little hard data in the literature.
- Dry granules have unlimited shelf life before being mixed with water.
- Short open time—working time may be less than 2 min unless parts are heated.
- Inconvenient—must be prepared and applied while hot, cools quickly.
- Limited life once mixed with water, extended with refrigeration.

6.1.2 Dynamics

One of the central concerns of a guitar designer is how to tailor the dynamic response of the body to produce a pleasing tone while still satisfying the competing constraints. There is yet no robust way to predict sound quality based on resonant frequencies and their associated mode shapes, but they are clearly important and worth learning a little more about.

In the early 1970s, Karl Stetson, a pioneer in the field of photomechanics, made some of the first double exposure holograms showing the mode shapes of guitar bodies. Figure 6.33 shows the first mode of a dreadnought guitar at 97 Hz. The two white tubes at the bottom are connected to a speaker and provide a very small force because of the acoustic pressure. Holography is sensitive to motions on the scale of a wavelength of laser light (less than 1 μm or 0.001 mm), so very little force is needed. Figure 6.34 shows the second mode shape of the same instrument at 205 Hz. Figure 6.35 shows the modes shape at 390 Hz. These historical images present some of the same information as the Chladni patterns and laser vibrometer images in the previous chapter.

There are several things to be learned from these images. The first is that the popular understanding of what the first few modes of an acoustic guitar look like are

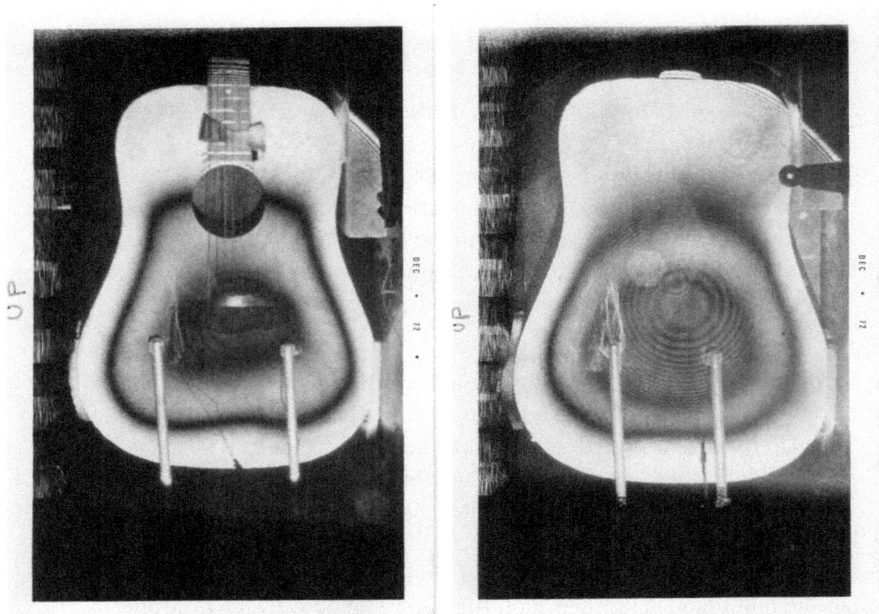

Fig. 6.33 Guitar top and back mode shapes, 97 Hz (images courtesy of Karl Stetson)

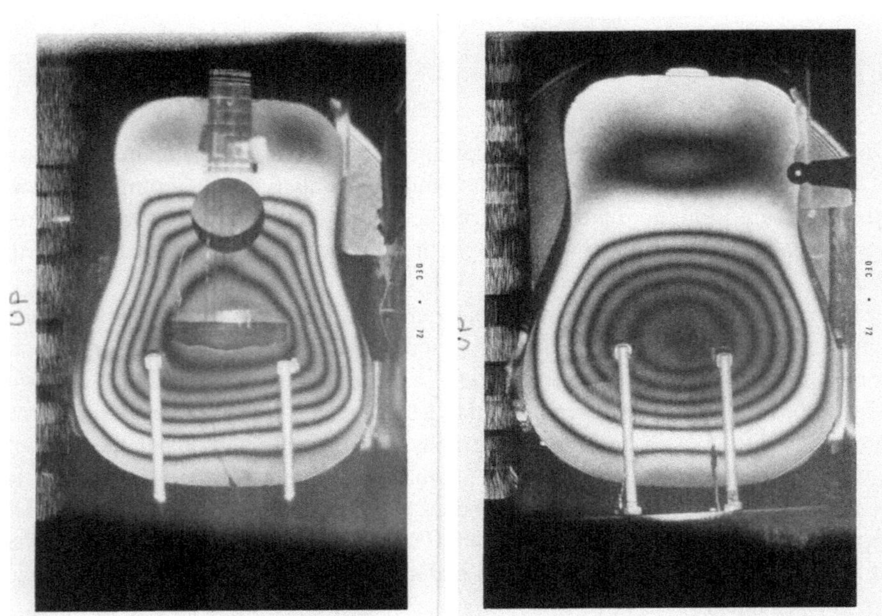

Fig. 6.34 Guitar top and back mode shapes, 205 Hz (images courtesy of Karl Stetson)

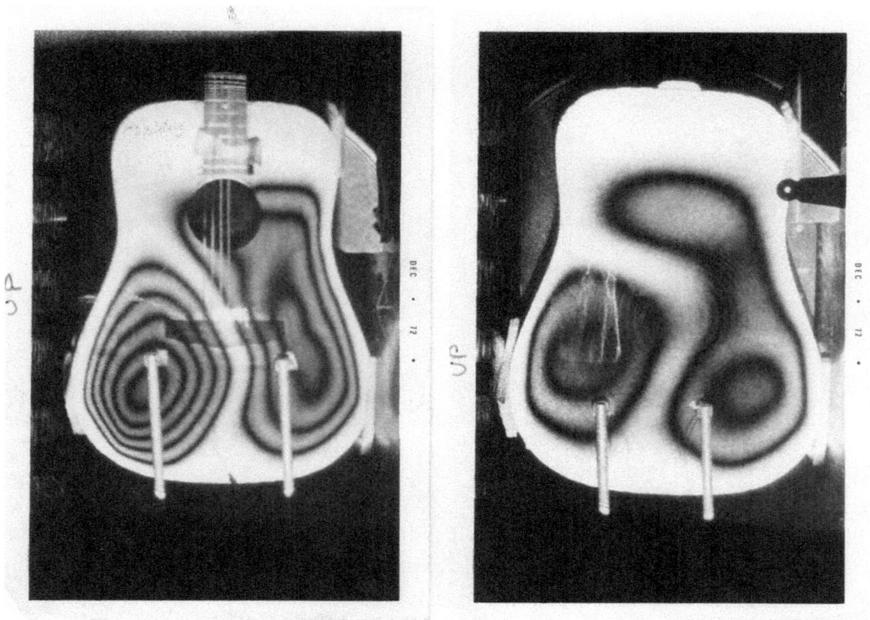

Fig. 6.35 Guitar top and back mode shapes, 390 Hz (images courtesy of Karl Stetson)

essentially right (see the previous chapter on Physics). The first two modes are often called monopole modes since there is a central peak in them and the top acts like a single speaker cone at these frequencies. The third mode at 390 Hz is a dipole mode since there are two peaks that are analogous to two closely spaced speaker cones. The two are out of phase with one another, so one moves up while the other one moves down.

The dynamics of acoustic guitars are complicated for several reasons. Material properties are usually known only approximately and the stiffness of wood is directional. Also, the geometry of guitars is hard to approximate with simple shapes. Finally, the strings, neck, body, enclosed air, and surrounding air are all coupled into a single system. As a result, detailed analysis requires sophisticated engineering software. Fortunately, simple and accessible approximations can give enough insight to be useful.

The dynamics of the neck are also quite important. The behavior of vibrating strings is strongly affected by the structure they are connected to. The familiar expression for the frequencies of an ideal vibrating string is

$$f_n = \frac{n}{2L}\sqrt{\frac{T}{\rho}}$$

Fig. 6.36 An idealized string test

where, n is the frequency number ($n = 1$ is the fundamental), L is length, T is tension, and ρ is mass per unit length. This expression assumes the ends of the string are connected to something rigid and massive, so they don't move at all and, so, is only approximately correct. Necks have both stiffness and mass, so they have their own resonant frequencies and their own impedance values. They are much heavier than the strings and have much more bending stiffness, but the dynamics of the neck can couple with the dynamics of the strings. Players sometimes find a neck that has a "dead spot," so a string fretted at that point doesn't ring well. That is, the dynamics of the neck are such that kinetic energy from the string is absorbed and the note decays quickly.

Before we move on, let's revisit the effects of damping in the structure of an acoustic guitar. Damping is simply a loss of energy in a vibrating structure. The energy doesn't magically disappear, but must go somewhere. In an acoustic guitar, the most important source of energy loss is radiated sound. The other two of most interest to luthiers are material damping and constrained layer damping from glue joints.

One important tonal characteristic of an acoustic guitar is its sustain—the time a note rings after being played. Some players would like to have a guitar that is both loud and has a long sustain. Unfortunately, these two goals oppose one another. The guitar can either radiate its energy quickly and loudly or parcel it out more slowly for longer sustain.

To illustrate the extreme case, I ran an idealized test to observe the vibration of a single string. A single plain string was mounted on a heavy ash block with ¼ in. (6.35 mm) beveled steel plates acting as the nut and saddle. The block was temporarily glued to a bench with a heavy ash top. The motion of the string was observed with a laser Doppler vibrometer, a device that measures velocity of the target without contacting it. It's akin to laser radar. Figure 6.36 shows the test setup, including a closeup of the string. The string was tapped with a lightweight instrumented hammer and the vibrations were allowed to decay.

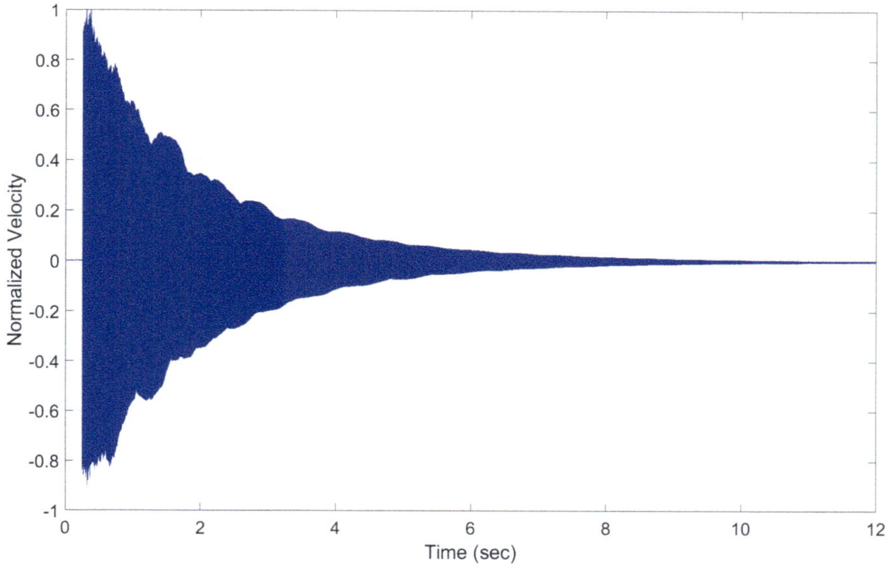

Fig. 6.37 Time domain response from idealized string test

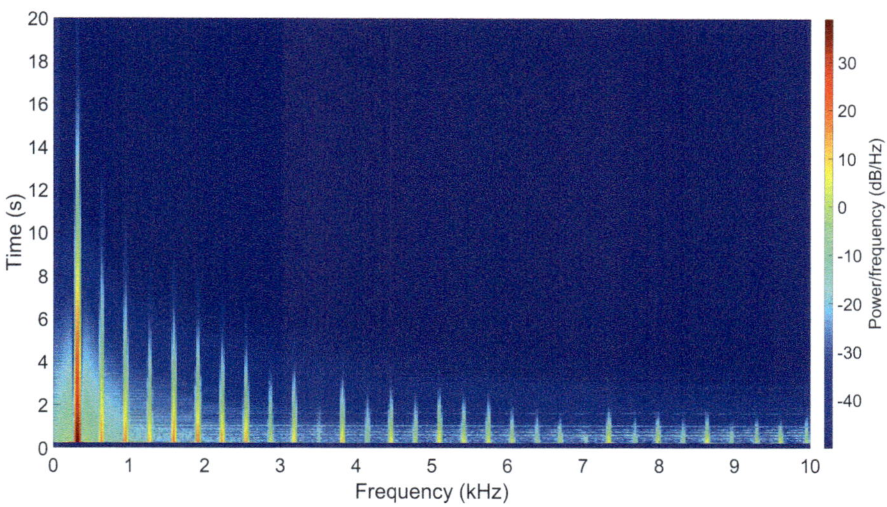

Fig. 6.38 Spectrogram from ideal string test

Figure 6.37 shows one of the signals recorded from the laser vibrometer. It takes more than 10 s for the response to decay completely.

A more informative way to look at this data is a spectrogram, a plot that shows both time and frequency, with frequency along the horizontal axis and time along the vertical axis. Colors show magnitude. Figure 6.38 shows the spectrogram from the idealized string test. Note that the fundamental was still observable nearly 20 s after the hammer impact. For comparison, pluck a string on an acoustic guitar and wait 20 s. Also, there were frequency components out to the maximum observable frequency of 22 kHz. This plot is limited to 10 kHz to make it easier to see the lower frequency components. One of the most important features of this plot is that contributions of the resonances drop off almost steadily in both time and frequency. This suggests that the dynamics of the fixture itself aren't contributing much to the response of the string.

For comparison, consider a recording made from a Taylor 710 dreadnought guitar. It was placed in a large hemi-anechoic chamber (cutoff frequency of 200 Hz) and the sound from the guitar was recorded using a binaural head, a very sensitive device that mimics human hearing. Figure 6.39 shows the spectrogram of the sound recorded from plucking the treble E string.

There are several interesting features in this spectrogram. To start, there is some background noise at low frequencies, unrelated to the sound made by the instrument. The chamber was very quiet and, for reasons having to do with the physics of sound waves, almost all of the noise was below 200 Hz. This noise appears as vertical patches of color at the far left of the plot. The string on the guitar doesn't ring as long as the ideal string because of energy lost to radiated sound and material damping. The sound made by the instrument depends on the relative contributions of the different frequencies and these vary. Note that the fundamental frequency (329.6 Hz) and the second harmonic have more power than the first harmonic. The third and fourth harmonic make only minor contributions to the overall sound.

To complete the picture, Fig. 6.40 shows a spectrogram for the bass E string of the same guitar. The horizontal range now stops at 5 kHz since the string has almost no content at frequencies higher than this. Similarly, the vertical range now stops at 10 s and there is no significant content after about 8 s. The relative contributions of

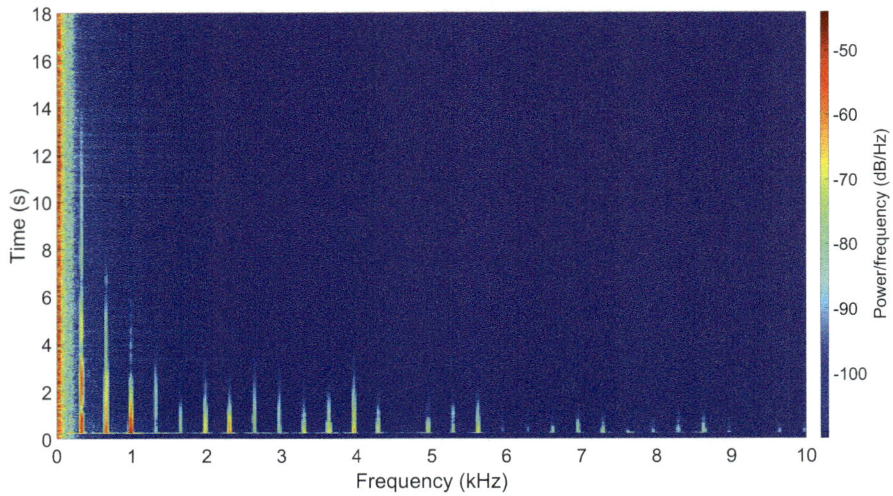

Fig. 6.39 Spectrogram of treble E string on Taylor 710 Guitar

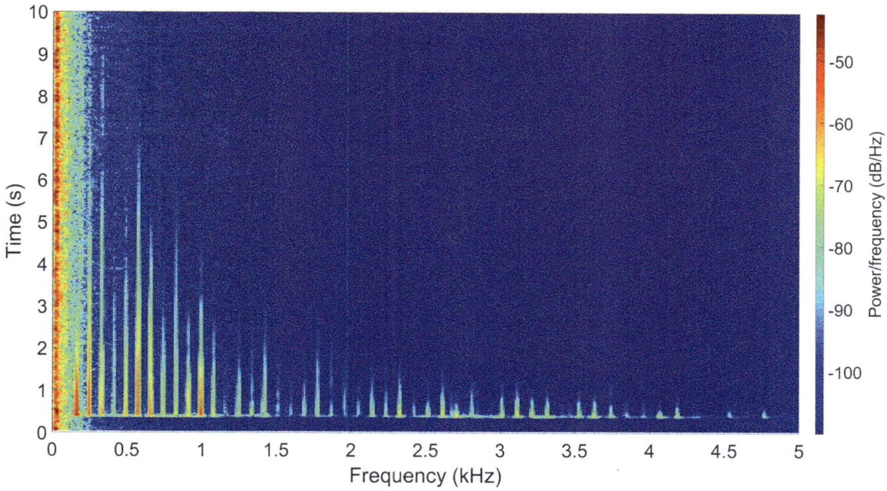

Fig. 6.40 Spectrogram of bass E string on Taylor 710 Guitar

the different frequencies are different than for the treble E string and this is part of what gives the instrument its unique sound. We should note that this is a very good guitar, having been rated favorably by numerous players.

Glue is potentially a source of damping in a guitar. If it is viscoelastic, it can act like the constrained layer dampers in cars. The physics of glue dictates that it is much stronger in shear than any other loading condition. In fact, standards for glued joints in aircraft dictate that they must be almost exclusively in shear, just as it would be in a constrained layer damper. For luthiers, a very common example is braces glued to the inside of a top. Figure 6.41 shows a small guitar with braces that were glued with AR.

The glue layer between a brace and the top experiences dynamic shear loads as the top vibrates, as shown in Fig. 6.42.

If the glue layer is basically elastic and the area of the hysteresis loop is small and little energy is lost in the shear layer. However, if the glue is viscoelastic, energy can be lost in deforming the glue. This energy loss is damping. Thin layers of hard-drying glues are assumed to add little damping.

Glue joints are strongest when the two surfaces fit tightly and the resulting glue layer is thin and even. The builder must create a tight, even joint, but not so tight that all the glue squeezes out. An ideal thin, even glue joint should not be able to add unacceptable damping. Using a hard-drying glue such as a harder formulation of AR glue or hot hide glue should add very little damping. Gore and Gilet observe that they know of no one who can identify the kind of glue used in a guitar just by listening to it.

It's difficult to quantify viscoelastic behavior in a material. However, in glue, the hardness of the dried glue is probably an acceptable proxy for how much damping it adds. Hot hide glue dries hard and adds little damping to a tight-fitting joint. In

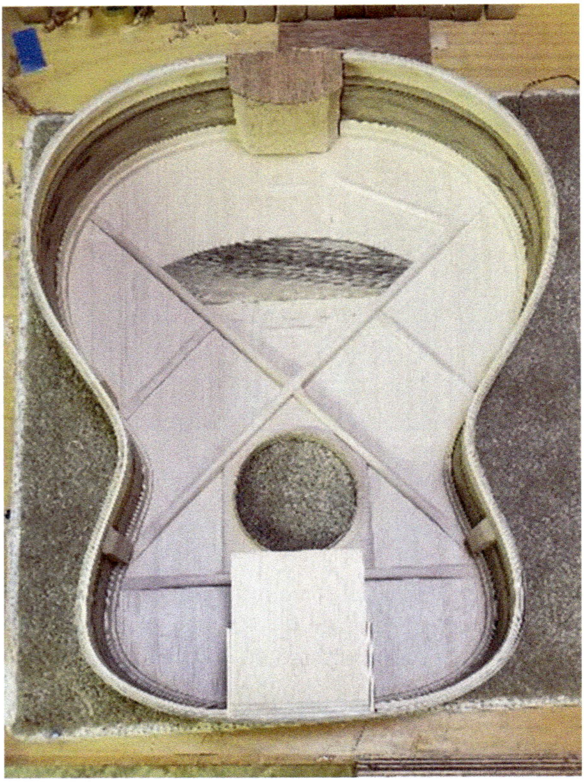

Fig. 6.41 A small guitar with X-bracing

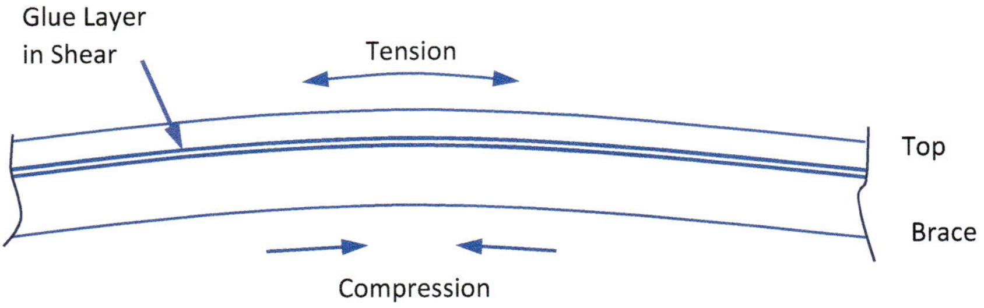

Fig. 6.42 A glued joint in shear

contrast, white polyvinyl acetate (PVA) glue, such as Elmer's Glue-All, doesn't completely harden and can be assumed to add significant damping to a glue joint.

Comparisons between AR and hide glues often rely on personal and shared experiences, though there are several formal engineering standards describing formulation and testing of glue. Two of the most relevant ones are ASTM D906 Standard Test Method for Strength Properties of Adhesive Bonds in Shear by Compressive Loading and ASTM D7966/D7966M-16 Standard Test Method for Resistance to

Creep of Adhesives in Static Shear by Compressive Loading (Wood-to-Wood). Manufacturers sometimes offer limited test data on their web sites.

Finally, the energy lost to damping has to go somewhere and in viscoelastic materials it turns into heat. There is so little kinetic energy in an acoustic guitar that the small amount of energy lost to damping doesn't make enough heat to measure. In the extreme, though, dampers can get hot. For example, the shock absorbers in motorcycle suspensions can absorb enough energy to get very hot. Some even have additional oil reservoirs to add heat capacity.

6.1.3 Back Structure

Most of the discussion around the structure of acoustic guitar bodies focuses on the top. However, there are braces on the back and they can affect the sound of the instrument. Luthiers have the choice of how to brace the back and most choose ladder bracing, often paying only minor attention to the resonant frequencies of the braced back plate. Even very rigid backs are dynamically coupled to the top and can contribute to the resonant frequencies of the instrument. A traditional ladder-braced back has heavy braces that can make the back stiffer than the top. Figure 6.43 shows such a back, with three heavy transverse braces.

Fig. 6.43 Traditional ladder-braced back

An alternative approach is popularly called a live back, in which the braces are designed to be more flexible and to couple more readily with the top. This is a bit of a misnomer since seems to imply that a ladder-braced back is rigid, but we are probably stuck with these names. Figure 6.44 shows a pair of live backs on guitars by Charles Fox. The three transverse braces are much lighter and the centers are significantly lower to increase flexibility. The transverse braces are joined by four radial braces.

6.1.4 Top and Back Radius

In spite of being popularly called flattops, steel string acoustic guitar tops and backs are seldom built flat. A slight curvature adds stiffness without adding mass and reduces distortion of the top from string tension. Since wood creeps slowly under load, permanent distortion can take years to become serious. Figure 6.45 shows this effect on a 1941 Martin 00-17. It's clear that the bridge has rotated forward, causing a dip, often called a belly, between the bridge and soundhole. The instrument was repaired. In fact, this image is from a video by Dan Erlewine showing the repair process.

Tops and backs on modern steel string guitars are generally built to a spherical curvature. That is, they are sections from a sphere of a given radius. It is typical for tops to have a longer radius than backs, though there is no standard. Top radii are

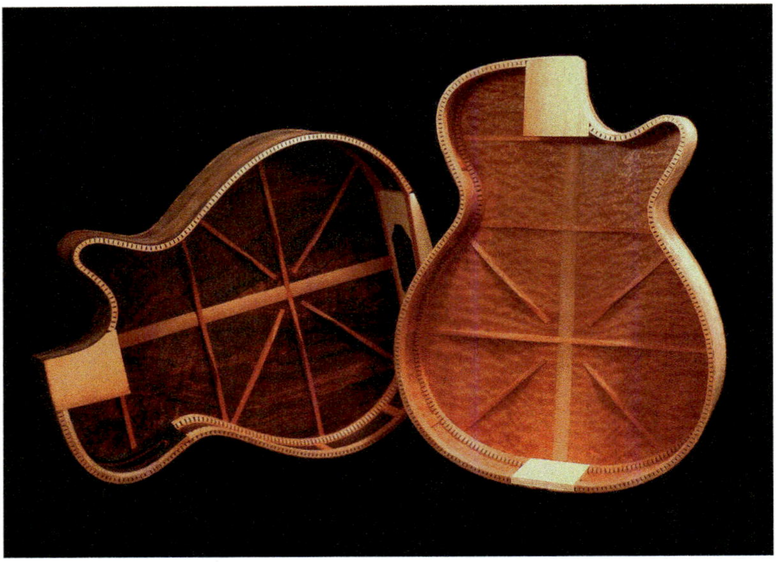

Fig. 6.44 Radially braced live backs (image courtesy of Charles Fox)

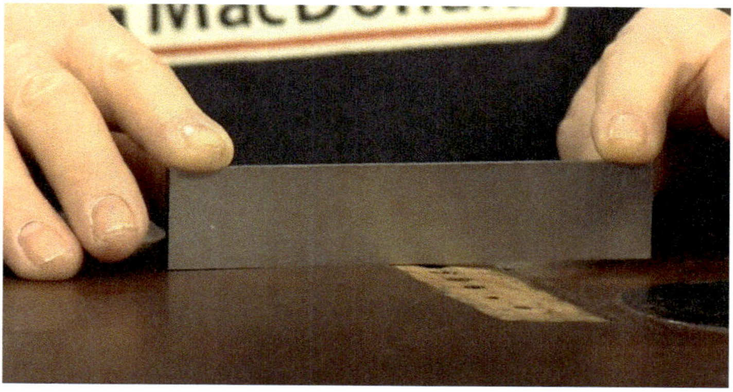

Fig. 6.45 The Warped Top on a 1941 Martin 00-17 Guitar (image courtesy of Stewart MacDonald, stewmac.com)

Fig. 6.46 A top being braced in a 25-foot radius dish

often in the range of 25–40 ft. (7.62–12.2 m). Back radii are often in the range of 12–25 ft. (3.66–7.62 m).

Steel string guitars are often built using dishes sanded to a specific radius. Figure 6.46 shows a top being braced on a 25 ft. radius dish during a class taught by Charles Fox. The back of this guitar was built to a 12 ft. radius. Since this guitar was made, Fox has increased top radius in this design to 30 ft. (9.14 m).

Fig. 6.47 A guitar body on a radius sanding dish

Figure 6.47 shows the rim of the same guitar on a 12-foot radius dish covered with a very coarse sanding disk. In this picture, the back of the rim is being sanded so that its edges conformed to the same sphere used by the back.

The back was glued on while being secured in a radius dish fitted with a vacuum port so that it was held at the right radius while the glue dried. Figure 6.48 shows such a dish in the author's shop, patterned after one designed by Charles Fox. This method helps ensure that the radii of the top and back are close to the desired values. Even using this integrated approach to holding the correct radius, tops and back do spring back a little so that the final curvature is a bit less, and the radius a bit more than the design calls for.

There is some debate about what top and back radii should be used. The top radius is usually larger than the back radius, so the top is flatter. For simplicity, some builders use the same radius for the top and back. I use a 25-foot (7.62 m) radius for both top and back since it prevents eager students from accidentally reversing the top and back radii.

Geometry figures into the choice of top radius. Radius adds stiffness that allows the top to be thinner and lighter while resisting the pull of the strings with acceptably small deformations. However, this radius means that the top is not perpendicular to the rim where it joins the neck. Neck angle, θ, varies but is not more than a few degrees (Fig. 6.49). On the high end of the range Bob Benedetto specifies $\theta = 4\frac{1}{2}°$ for a full-sized archtop with a floating bridge. On the low end of the range, Don MacRostie specifies $\theta = 3/4°$ for his plan of a 1933 Martin OM. This guitar has a top radius in the neighborhood of 40 ft. (12.2 m), so it is almost flat.

The increased stiffness due to curvature also increases the resonant frequency of the top, since it doesn't add corresponding mass. A published analysis of a 0.5 m (19.68 in.) square plate with an initial curvature shows how the amplitude of the curvature affects the first resonant frequency (Fig. 6.50). The plate is about the size

Fig. 6.48 A vacuum radius dish for a guitar back

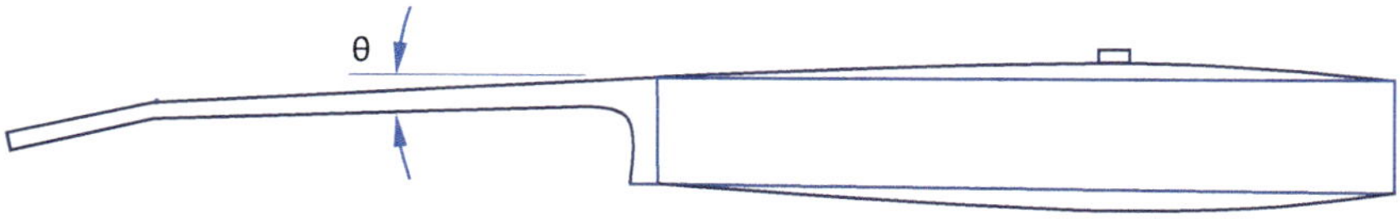

Fig. 6.49 Neck angle on a steel string guitar

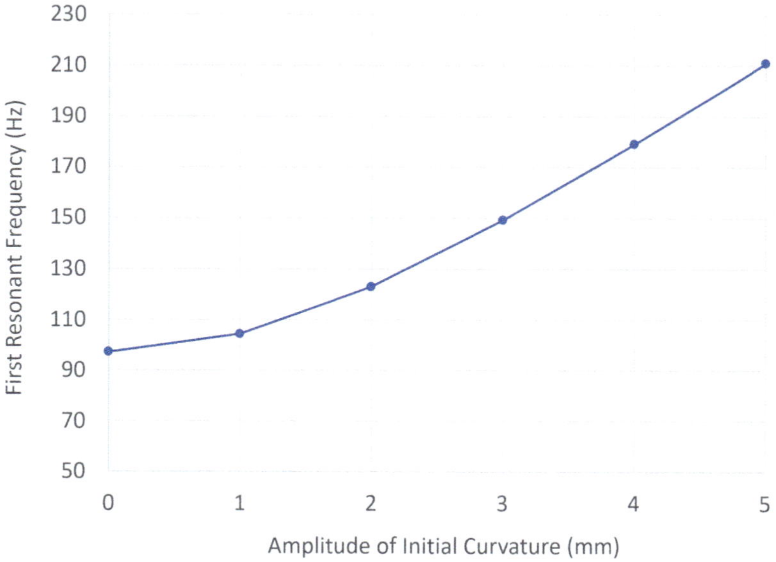

Fig. 6.50 Effect of curvature on first resonant frequency of a rectangular plate (analysis by Yu, Mostaghel and Fu)

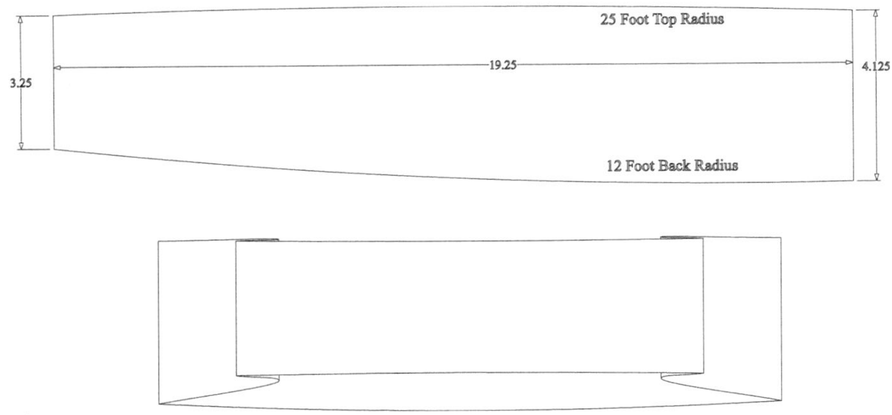

Fig. 6.51 Side and front view of a rim with radiused top and back

of a guitar top. It is assumed to be made of metal, so it is only analogous, but the trend is clear. The additional stiffness added by curvature can significantly increase the resonant frequency.

A final note about spherical tops is that the top and bottom edges of the rim cannot be flat if they are sanded to match the top and back radii. For convenience, acoustic guitars are sometimes sketched from the side as if the top and bottom of the rims are flat, as in Fig. 6.49. However, this isn't quite right. Once the completed rim is sanded in a spherical radius dish, the only straight lines on it are vertical (Fig. 6.51). When assembling the instrument, it's important to have a reference surface, preferably one that is straight in at least one direction. Once the rim has been radius sanded, the surface to which the heel joins can be used as a reference surface.

Classical guitars are traditionally built in a form called a solera and don't usually have a spherical top or back. This is clearly shown in a simplified solera by Paco Chorobo (Fig. 6.52). In this guitar, the upper bout is flat, but the lower bout isn't. The opening at the lower bout accommodates the portion of the top that is domed. The maximum height at the center is set by a small stack of veneer placed at the center. The cams around the edge hold the sides against the top while the instrument is assembled, a slight variation on traditional designs. Note that the outline of the body is flat, making assembly in the solera easier.

Figure 6.53 shows a top being braced in this solera. Note that the two lateral harmonic bars are flat. The five fan braces were not radiused before being glued on. Rather, they were just pressed into the solera, naturally taking on a curved shape that was fixed once the glue dried. Because the top and fan braces were sprung into place, there was some spring back when the top was removed from the solera. Note that the sides were glued to the edge of the top rather than the inside face. This allowed the cams to press the sides against the top. In this approach, the shape of the top defines the shape of the instrument.

Traditional soleras are heavily built and usually have removable sides. They are usually flat around the edge and have a flat upper bout, with a dish sanded in the

Fig. 6.52 A simplified solera made by Paco Chorobo

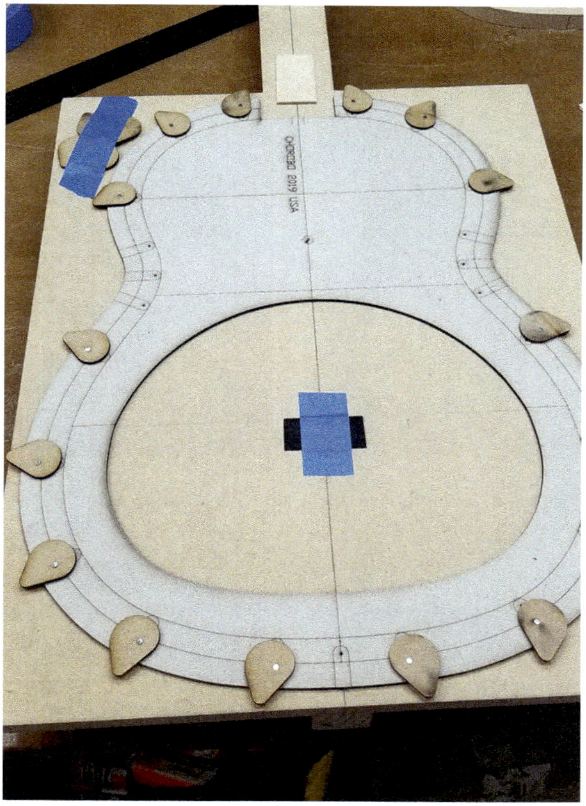

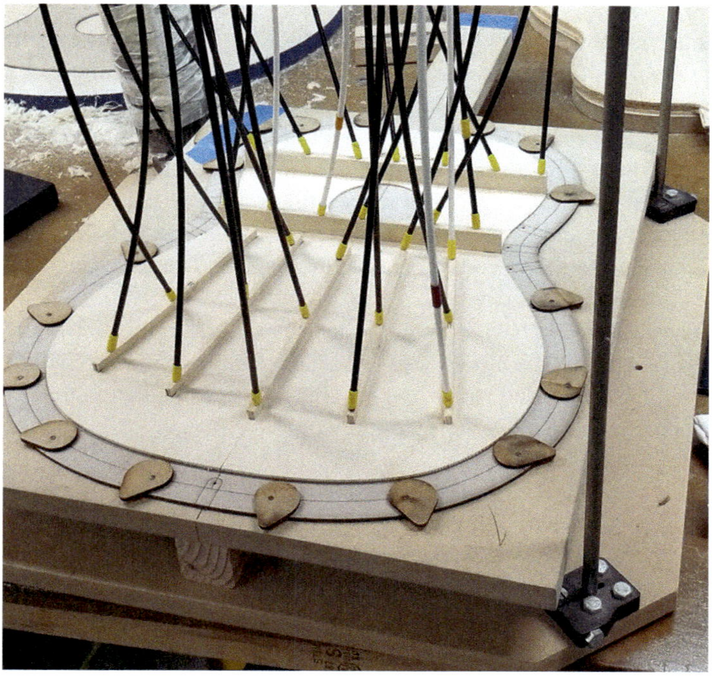

Fig. 6.53 Top being braced in a simplified solera

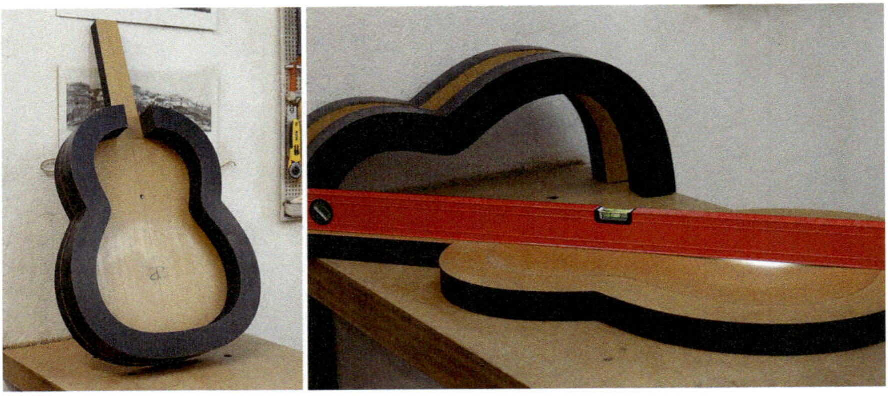

Fig. 6.54 A traditional solera showing concave lower bout (image courtesy of Paco Chorobo, chorobo.com)

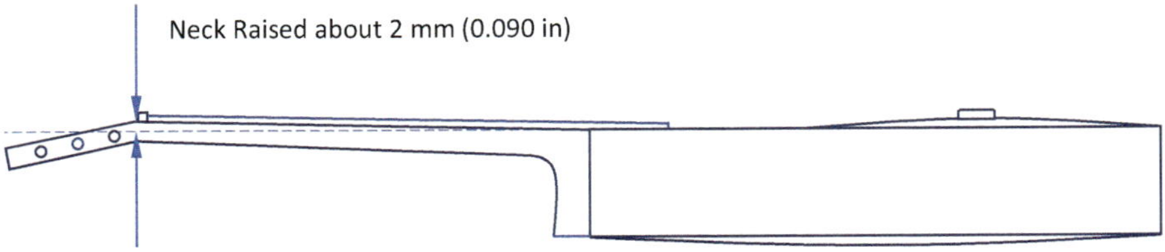

Fig. 6.55 Side view of representative classical guitar

lower bout. Figure 6.54 shows a solera made by Paco Chorobo. The picture on the right clearly shows the concave lower bout. A few luthiers use a more complicated geometry in which the seam between the sides and top is not flat.

While most classical guitars have domed lower bouts, it's an oversimplification to assume that they were all made that way. They are sometimes built with flat tops and flat bridges—the Ramirez 1a is a successful example. Richard Bruné notes that some were built flat and a few even used inverted domes. He observed that Herman Hauser II made guitars that were concave at the waist and that Domingo Esteso used a mixture of dome and concavity.

Most classical guitars have the end of the neck raised slightly above the plane of the body (Fig. 6.55). This allows a lower saddle on the domed soundboard while still maintaining the needed string height. This requires that the bottom of the fretboard be tapered where it overlaps the top.

6.1.5 Soundhole

The soundhole in an acoustic guitar has a large effect on the sound at lower frequencies. It affects the first two coupled modes of the body and also radiates sound itself. Most guitars have a single soundhole, so designers need to decide how large to

Table 6.1 Soundhole sizes

Guitar	Scale length (in.)	Scale length (mm)	Diameter (in.)	Diameter (mm)
Cordoba 24T (tenor ukulele)	17	431.8	2.44	62
Cordoba Mini	20.08	510	2.99	76
Martin Size 5 (Terz)	21.4 or 22	543.6 or 558.8	3¼ (3.25)	82.6
Cordoba Hauser	25.59	650	3.31	84
1937 Hauser	25.59	650	3.39	86
Martin Size 0	24.9	632.5	3 5/8 (3.625)	92.1
Martin Size 00	24.9	632.5	3¾ (3.75)	95.3
Martin Size 000	24.9	632.5	3 7/8 (3.875)	98.4
Martin M (0000)	25.4	645.2	4	101.6

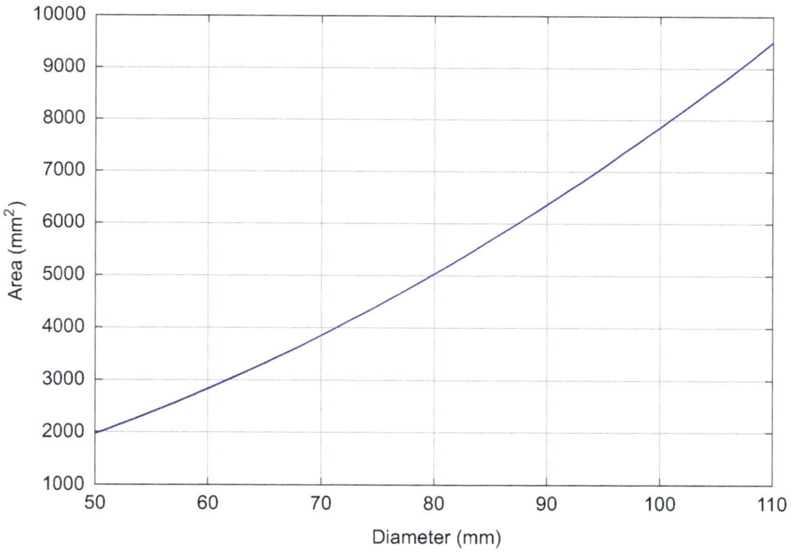

Fig. 6.56 Soundhole area and diameter (mm)

make it and where to put it. Archtops typically have two soundholes, one on each side of the bridge. A few instruments have a pattern of smaller soundholes, with Ovation guitars being the most well-known.

The range of practical sizes for single soundholes is small. Table 6.1 lists soundhole diameters for a selection of different sized instruments.

The most important characteristic of a soundhole is area. Area is

$$A = \pi r^2 = \frac{\pi}{4} d^2$$

where π is approximately 3.14159, r is radius, and d is diameter.

It's easier to look at plots. Figure 6.56 shows area vs. diameter in mm^2 and Fig. 6.57 shows the same relation in in^2.

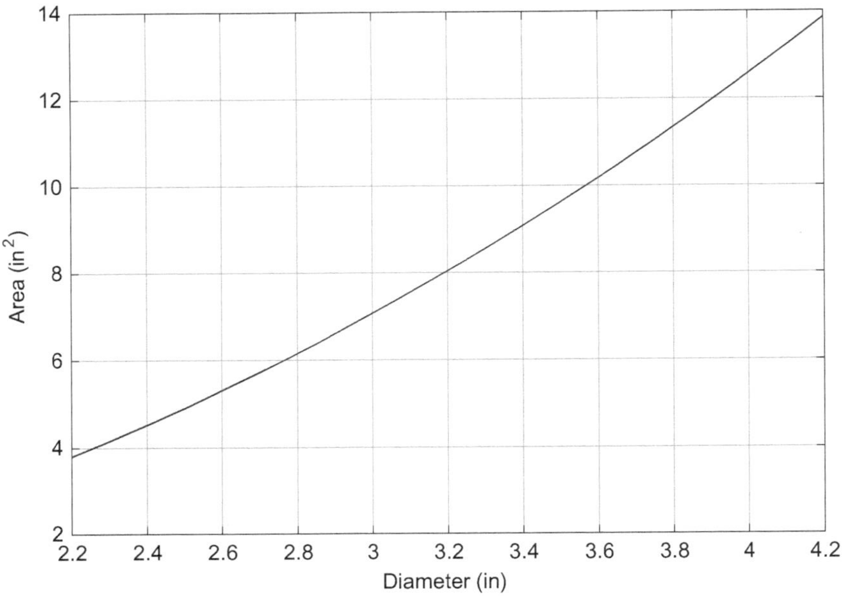

Fig. 6.57 Soundhole area and diameter (in.)

The size of the soundhole strongly affects the first resonant frequency and the Helmholtz frequency, with a lesser effect on higher modes. In particular, reducing the area of the soundhole tends to decrease the first coupled mode (breathing mode) and the Helmholtz frequency. However, it also reduces the area available to radiate sound at that frequency.

Another design choice is where to place the soundhole. By far, most are placed on the centerline, but this causes a structural problem, as it puts a large hole in the load path through the top. Some designs solve this problem with an offset soundhole, as shown in Fig. 6.58. This top is by McPherson, a manufacturer of both wood and carbon fiber guitars whose signature feature is an offset soundhole. Moving the soundhole gives more design freedom with the bracing. This top uses a modified X-brace with several interesting design features. The braces are laminated and the X-braces pass through each other without touching.

A secondary purpose of the soundhole is access to the inside of the guitar. For this, the soundhole must be large enough to accommodate a hand and 3 in. (76.2 mm) is about the lower practical limit. For reference, the smallest soundhole I've been able to fit my own hand through is on a Fender MA-1 ¾ size guitar, with a diameter of 3.3 in. (83.8 mm).

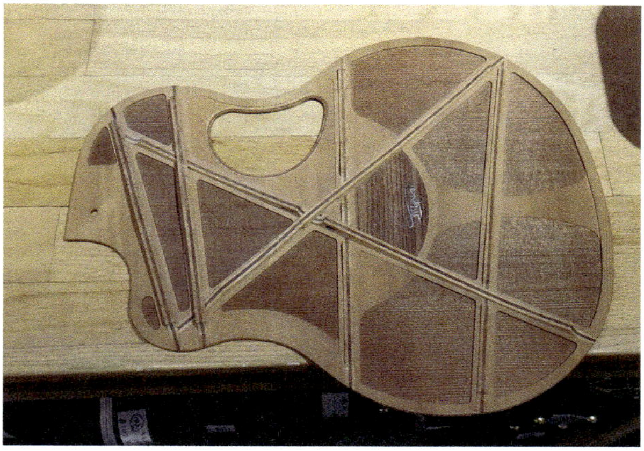

Fig. 6.58 The top of a MacPherson guitar with an offset soundhole (image by author, reproduced courtesy of The Music Gallery, musicgalleryinc.com)

6.1.6 Bridge

Acoustic guitar bridges are at once one of the most visible elements of acoustic guitar and the often most overlooked. On steel string acoustic guitar, bridge shapes are often proprietary and can serve as logos for a brand. However, bridges are also very important structural elements that add significant mass and stiffness to the top. Thus, the bridge can be one of the most important aesthetic and structural components of an acoustic guitar.

Acoustic bridges are almost always one of three types: fixed steel string bridges, fixed classical bridges, and floating bridges. Fixed bridges are glued directly to the top and floating bridges are held in place only by the downward force of the strings passing over them.

Classical guitars almost all use some variation of a simple rectangular fixed bridge with a saddle and tie block. Fig. 6.59 shows a traditional bridge on a classical guitar by Jeff Elliott. The strings are tied onto the block behind the saddle using self-locking knots.

Some builders have noted that this design limits the break angle over the saddle because the loop over the tie block lifts the string slightly. A solution is to add a second or even third set of holes so that the strings can be tried securely without lifting them as they approach the tie block. Figure 6.60 shows a 12-hole tie block on a classical guitar made by Angela Waltner. The additional set of holes allows the string to pass through the tie block without being lifted by the loop that secures it.

The path of the string through the doubled set of holes is shown in Fig. 6.61.

Fig. 6.59 A traditional classical guitar bridge (image courtesy of Jeff Elliott, elliottguitars.com)

Fig. 6.60 A 12-hole tie block (image courtesy Guitar Salon International, guitarsalon.com)

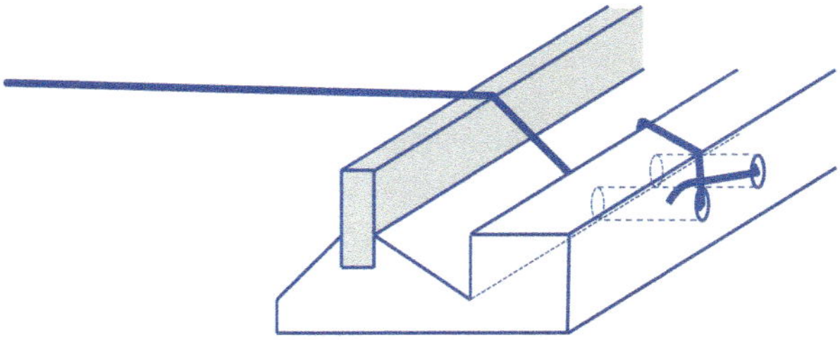

Fig. 6.61 String path through a doubled set of tie block holes

Fig. 6.62 A classical guitar bridge with three holes in the tie block per string (image courtesy of Paco Chorobo, Chorobo.com)

Fig. 6.63 A fixed bridge with pins on a 1954 Martin OO-18 (image courtesy of Mike and Mike's Guitar Bar, mmguitarbar.com)

A less common design uses three holes per string, as shown in Fig. 6.62. While a little more complicated to make, the strings are extremely well secured.

There is debate about the role of break angle over the saddle in classical guitars and Richard Bruné notes that good ones don't always have large break angles. He also notes that drilling additional holes in tie blocks can weaken them, causing splitting, something he never sees with six-hole tie blocks.

Fixed steel string bridges usually use pins to secure ball-ended strings, though some are pinless. Probably the most familiar fixed steel string bridge is the "belly bridge" used by Martin Guitars, starting in the 1930s. Figure 6.63 shows this bridge on a 1954 Martin 00-18. The saddle fits tightly into a straight slot that is slightly angled to improve intonation.

Figure 6.64 shows a Martin D-21 with the older pyramid bridge design and bridge pins with decorative contrasting dots.

Fig. 6.64 A pyramid bridge on a modern Martin D-21 (image courtesy of Mike and Mike's Guitar Bar, mmguitarbar.com)

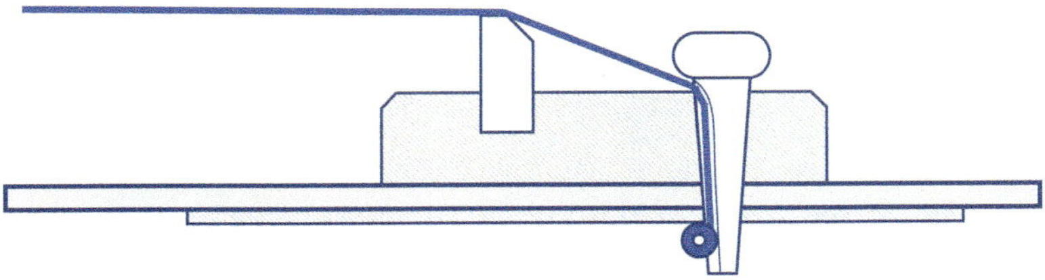

Fig. 6.65 String secured by bridge pin

It's worth a moment to review what the bridge pins do and why they are needed in this type of bridge. Figure 6.65 shows a cross section of a bridge with the pin in place. The pin prevents the ball on the end of the string from pulling back through the hole in bridge. The ball should pull up against the bridge plate, but should not wear the edge of the hole through the bridge plate.

Hardly anything in the guitar world is standardized and bridge pins are no different. Some have a 5° taper and some have 3° taper (Fig. 6.66). At this writing, there is no consensus on which are more popular, though one major parts vendor stocks 5° pins almost exclusively.

Figure 6.67 shows the dimensions of 5° bridge pins from Waverly. Builders should note that oversize pins are also available from some suppliers in case a reaming tool went too far into the bridge or an older bridge is seriously worn.

The pins themselves are commonly made of many different materials, including wood, bone, a variety of plastics, and even some metals. At this writing, D'Addario even offers titanium bridge pins, as shown in Fig. 6.68. Note the chamfered ends to reduce the possibility of the ball jamming under the end of the bridge pin and not seating correctly against the bridge plate.

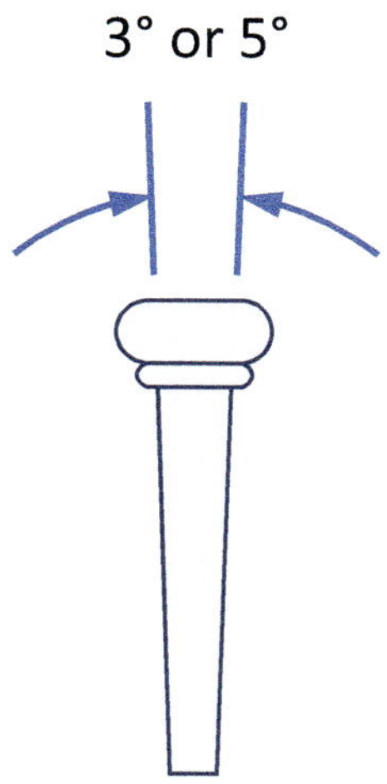

Fig. 6.66 Taper angle of bridge pins

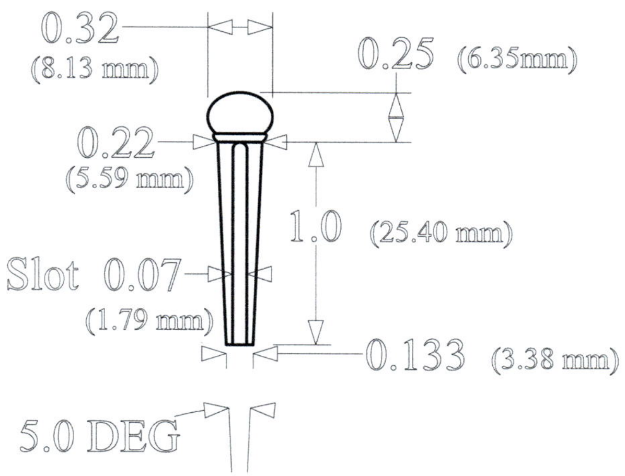

Fig. 6.67 Dimensions of 5° taper Waverly bridge pins (from dimensions provided by Stewart MacDonald, stewmac.com)

Fig. 6.68 Titanium bridge pins (image courtesy of D'Addario, daddario.com)

Fig. 6.69 A Taylor 114e showing the distinctive Taylor bridge shape

Vintage guitars sometimes have ivory bridge pins, but these are fortunately no longer used. At this writing, at least one vendor offers fossilized mammoth ivory bridge pins. They are predictably expensive and one wonders how much of the raw material must be available. Still, they are ecologically preferable to ivory taken from living animals.

It's not unusual for the bridge shape to serve as a way of identifying a specific brand. One of the most familiar is that used by Taylor Guitars. Figure 6.69 shows a Taylor 114e with the distinctive Taylor bridge shape.

Not all steel string bridges use pins to secure the strings. Though far from universal, pinless bridges are not uncommon. The Ovation Balladeer, introduced in 1966, had a pinless bridge, so they date from at least then. Most new things in the guitar world have old roots, so the concept may be much older. Figure 6.70 shows the bridge on a 2006 Ovation Balladeer.

This design has several advantages over the more common pinned bridge. The obvious one is that the strings are easier to change. Another is that there doesn't need to be a bridge plate on the inside of the relatively soft soundboard to protect it from the ball ends of the strings. This offers the designer more latitude in placing the braces.

If there is a disadvantage, it's likely that the bridge is completely dependent on the glue joint for its integrity. A pinned bridge experiences a force between the

Fig. 6.70 A pinless bridge on an Ovation balladeer (image courtesy of Mike and Mike's Guitar Bar, mmguitarbar.com)

Fig. 6.71 The bridge on a Breedlove 12 string jumbo guitar (image courtesy of Mike and Mike's Guitar Bar, mmguitarbar.com)

bridge and soundboard due to the tension of the strings and that force is missing with a pinless bridge. The shear strength of glue is measured in thousands of psi and tens of millions of Pa (1000 psi = 6.9 MPa), so a glue area of the size of a postage stamp is enough to resist string tension in the ideal. However, guitars seldom inhabit an ideal world and are sometimes treated roughly. Still, there doesn't seem to be a history of Ovation bridges coming off.

A particularly interesting pinless bridge is one shown in Fig. 6.71 on a Breedlove 12 string guitar. It has a very distinctive design and securely anchors all 12 strings, which have nearly twice the tension as a set of six strings. It apparently holds just fine.

Fig. 6.72 A split saddle on a Takamine Dreadnought guitar

It's rare for acoustic guitar bridges to have individually adjustable saddles. The additional weight is problematic since acoustic tops need to be light in order to respond to string vibrations. By far, most acoustic saddles are one piece and have a slight angle to improve intonation. Additional improvements in intonation come from slightly altering the top profile at each string. It's common for the B string of a steel string guitar to require more saddle offset than a simple straight saddle can accommodate. Thus, many saddles, such as the one shown on the Ovation Balladeer in Fig. 6.70, have a discontinuous ledge molded in.

Another approach is to split the saddle into two pieces to allow for more flexibility in intonation.

Figure 6.72 shows a Takamine Dreadnought guitar with a split saddle on a pinless bridge. This guitar is owned by a musician whose day job is as a middle school science teacher. I saw him playing this well-loved instrument in a steak house in the mountains overlooking Phoenix. Alas, I failed to get his name. If you're out there, sir, thanks for letting me take this picture.

An ideal bridge would have adjustable saddles as do most electric guitar bridges. The obvious problem is that acoustic bridges need to be light, so a heavy adjustment mechanism is not acceptable. A few builders have experimented with adjustable saddles. A nice example is by Jay Dickinson of Portland Guitar Co. (Fig. 6.73). Dickenson has also developed an adjustable nut that works in a similar way. Dickinson holds two patents on his adjustable nut and saddle, but was at least one previous patent. Manuel Ferdinand Rodriguez was granted Spanish patent ES108654U in 1965 for a bridge with adjustable saddles. That patent has since expired.

It is not uncommon to fit bridges to uncompleted instruments temporarily using hollow bolts inserted into the bridge pin holes, as shown in Fig. 6.74. They are

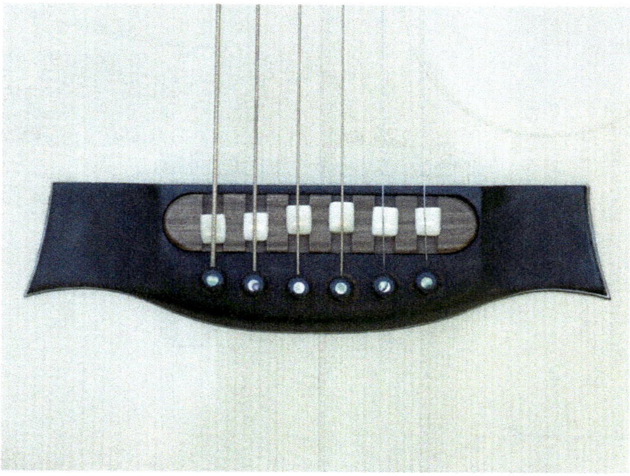

Fig. 6.73 An acoustic bridge with adjustable saddles (image courtesy of Portland Guitar Co., portlandguitar.com)

Fig. 6.74 A bridge fitted temporarily with hollow bolts

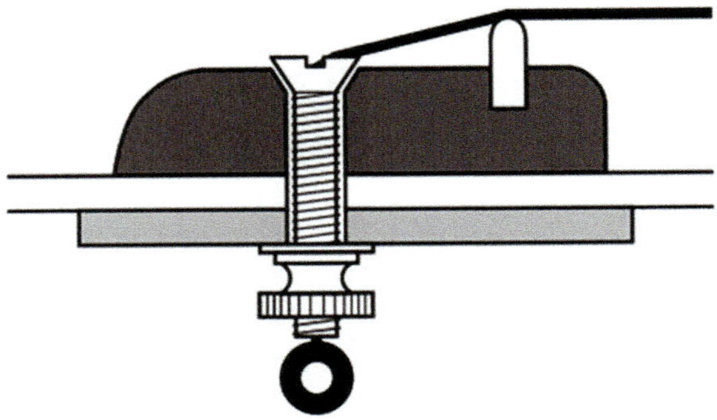

Fig. 6.75 Strings passing through hollow bolts (image courtesy of Stewart MacDonald, stewmac.com)

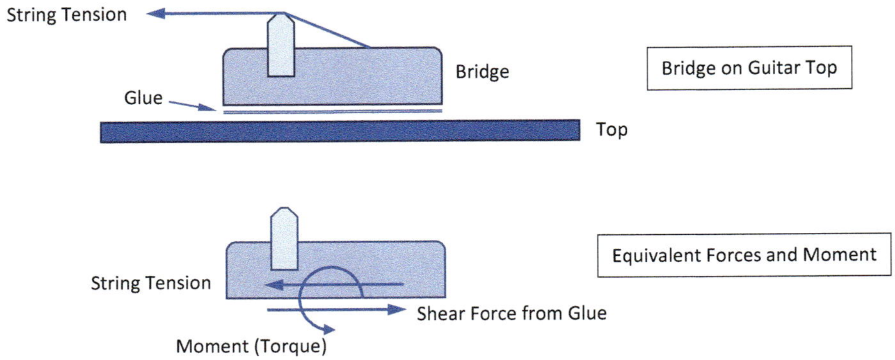

Fig. 6.76 Forces acting on a pinless fixed bridge

secured on the inside of the instrument with nuts and washers while the strings are inserted from the inside, as shown in Fig. 6.75. The instrument shown in Fig. 6.74 was made by the author as part of a class taught by Charles Fox. The class schedule did not include time for finishing, so we bolted the bridges on temporarily and strung them so we could set them up. The instrument sounded good, in spite of the additional weight from the bolts.

The forces acting on a fixed bridge are easy to understand at a basic level. The strings apply a horizontal load above the glue layer between the bridge and top, as shown in the pinless bridge in Fig. 6.76. Using basic ideas from structural mechanics, the string force can be resolved into a horizontal force at the glue layer and a torque (engineers call it a moment). This could be either a pinless steel string bridge or a classical bridge.

The equivalent force and moment show that there is a shear force across the glue layer and a moment that tries to lift the back of the bridge. This matches experience since bridges tend to lift from the back when they come loose.

The story is a little different for a bridge that uses pins to secure the strings as shown in Fig. 6.77. The ball end of the string creates a downward force on the bridge that is absent on the pinless bridge.

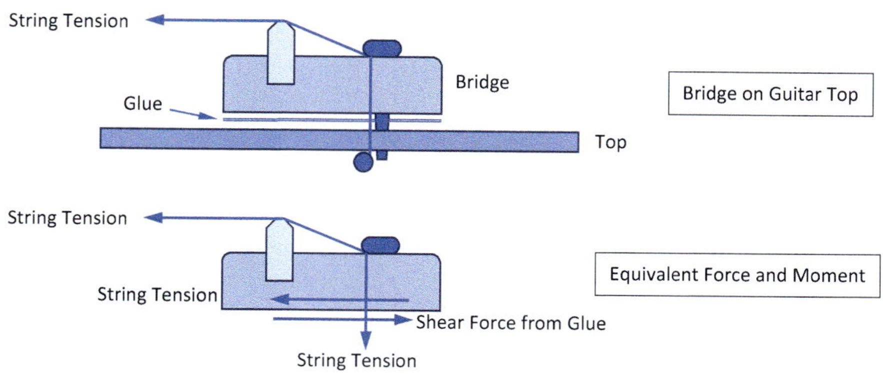

Fig. 6.77 Forces acting on a pinned bridge

The next step is to compare the stresses with the strength of the most commonly used glues. Yellow wood glue, such as Titebond, has been widely used in the guitar industry for decades and is nearly universal, though classical builders sometimes still use hot hide glue. Titebond has a listed shear strength of 3600 psi (24.8 MPa). Let's assume a typical bridge has a gluing area of 6 in.^{2} (3871 mm^2) and tension of a heavy set of steel strings is 215 lb. (956 N which is equivalent to 97.5 kg). The shear stress on the glue layer holding the bridge to the top is then 35.8 psi (0.25 MPa). Thus, the stress on the glue joint is tiny compared to the nominal strength of the glue. Hot hide glue is about as strong as wood glue, so the conclusion is the same. If this is true and glue is far stronger than it needs to be in order to bear string tension, then why do bridges ever lift?

For pinless bridges, the applied moment tends to lift the back of the bridge and glue is weaker in tension than it is in shear. Another reason is that bridges almost never fail exactly at the glue joint. Rather, there is often some wood lifted as well. Finally, the listed shear strength of glue is under the most ideal conditions. Glue has the most shear strength when used in a thin, even layer. It is generally tested at ideal room temperature and humidity and weakens in high temperatures or high humidity conditions.

The grain of the bridge is perpendicular to that of the top, so extremely high or low humidity can create stresses in the glue joint as the two try to expand or contract. These stresses are added to those due to string tension. It's never a good idea to expose a wood guitar to extremes of humidity.

Since the pinned bridge doesn't have a moment that would cause it to lift, the failure mode is different than for pinless bridges. Damage to the bridge plate can allow the ball ends to pull through. When this happens, the vertical force provided by the ball end of the string against the bridge plate is then borne by the bridge and the forces look like those on a pinless bridge.

Fig. 6.78 A classical style bridge fitted with steel strings (Wikimedia Commons, commons.wikimedia.org)

Perhaps the most common problem in bridges is poorly fitted joints. While it is certainly possible to fill a gap with the glue, the resulting joint is weak. The ideal joint should, according to Paco Chorobo, "fit like glass on glass." Tight, clean joints are not just the mark of a skilled builder, they are naturally stronger.

All this supports the conclusion that a well-designed and well-fitted bridge that is secured by either hide glue or an appropriate modern wood glue should be able to bear any reasonable string load in the absence of mistreatment. Accordingly, Fig. 6.78 shows a classical bridge fitted with steel strings. This is unusual and is generally avoided since classical guitars are not designed for the higher tension of steel strings. However, the guitar in this picture appears to have a pick guard, suggesting that it was designed for steel strings.

Not all bridges are permanently mounted to the top of the instrument. Floating bridges are moveable and held in place only by the tension of the strings. They are most widely used on archtop guitars, such as the vintage Kay K1-B shown in Fig. 6.79. In this design, the bridge bears almost no loads in the direction of the strings. Rather, there is a normal force due to the break angle of the strings over the bridge. This, in turn, puts the entire body in tension. Thus, the bracing must bear vertical loads that are absent in guitars with fixed bridges.

Note that flattop instruments also occasionally have floating bridges. Figure 6.80 shows a 1967 Stella Harmony six-string guitar. Externally, the only clear difference between this instrument and one with a fixed bridge is the tailpiece. This guitar is an example of a long line of inexpensive instruments by Harmony and Stella that often had stamped steel tailpieces. Stella was a brand owned by the Oscar Schmidt Company. Stella was acquired by Harmony in 1939 and the brand was eliminated in 1974. The names Stella and Harmony are often used together.

Another flattop acoustic guitar known for having floating bridges were introduced by the French company Selmer in partnership with guitarist and luthier Mario

Fig. 6.79 A 1950s Kay K1-B archtop with a floating bridge (image courtesy of Mike and Mike's Guitar Bar, mmguitarbar.com)

Fig. 6.80 A 1967 Stella Harmony (image courtesy of Yooptone Music, yooptone.com)

Maccaferi (Fig. 6.81). This distinctive style of instrument was made famous by jazz guitarist Django Reinhardt.

Fixed bridges are major structural elements since they can add both stiffness and mass to the top. Particularly, on a classical guitar, it may be the largest single

Fig. 6.81 A 2018 reproduction of an O-Type Selmer-Maccaferri by Jürgen Lutschkowski (Wikimedia Commons, commons.wikimedia.org)

contribution to the stiffness of the top, particularly in the lateral direction. Thus, it can be an oversight to focus on the top bracing while paying only passing attention to the contribution from the bridge.

Bridges are usually made from hard, heavy woods such as ebony and rosewood. Production guitars sometimes use alternative woods, dyed to match the look of ebony. The ideal bridge is light, stiff, and hard enough to form a good seat for the saddles. Unfortunately, no wood approaches this ideal. Some luthiers reinforce the wood bridge with other materials such as carbon fiber. Trevor Gore states that carbon fiber reinforcement allows alternative woods to be used, with the result being up to a 50% decrease in the mass of the bridge. Lighter tops respond more quickly to the vibration of the strings.

Fig. 6.82 Bridge plate with angled grain

A fixed bridge acts as a heavy lateral brace across the top. A heavy bridge for a steel string guitar is much more rigid than the braced top to which it is glued. It may be effectively rigid compared to the portion of the top it covers. This is compounded if there is a heavy bridge plate.

The bridge plate needed for a pinned fixed bridge necessarily adds mass and stiffness. Less obvious, though is the fact that it can add a source of structural failure. The bridge plate needs to have a row of holes in it for the bridge pins. If these holes line up with the grain of the bridge plate, they can cause it to crack. A simple refinement is to make sure the grain of the bridge plate is not parallel to the row of holes. Figure 6.82 shows a bridge plate with the grain angled in this way. It is typical for builders to cut bridge plates from scraps left over from the back or sides. As one might guess from this bridge plate, the back and sides are made from East Indian rosewood.

6.2 Neck

The neck is a major structural element and it's quite important to get it right. The basic needs are strength and alignment. Almost any neck is strong enough to bear string loads, particularly if it includes one or more truss rods. Also, the neck needs to be precisely aligned with the body so that string heights are reasonable. This is a routine part of just about any build.

The difference between a good neck and a great one is often subtle. Four less obvious features are:

- Neck placement
- Neck dynamics
- Shape
- Weight

Fig. 6.83 Christopher Morrongiello playing a baroque guitar (Wikipedia Commons, commons.wikimedia.org)

The designer needs to decide on how to place the neck with respect to the body. The earliest guitars had the fretboard flush with the top, to the point that higher frets were set directly into the soundboard. Figure 6.83 shows a modern reproduction of a baroque guitar made this way. While it certainly works—instruments were made this way for hundreds of years—it can be problematic for both the builder and the player. The neck is necessarily part of the body so resetting the neck would be very difficult. Also, access to the higher frets could be inconvenient.

More modern acoustic guitars almost all have the fretboard raised above the body so that the glue seam between the fretboard and neck is even with the top. This accommodates both glued and bolted neck joints. It also accommodates body cutaways that are the almost universal solution for improved access to higher frets.

Another, more involved way of improving neck access is to raise the neck above the plane of the soundboard. Raised fretboards occasionally appear on both classical and steel stringed instruments. This is not a new idea, but seems to be the subject of renewed interest. While attractive to many players, it can complicate the joint between the neck and body. Figure 6.84 shows an Ergo model guitar by Charles Fox. This instrument includes a range of design refinements, including a raised neck. Also visible in this picture are the wedge body, a sound port with door, and lightening holes on the transverse bar below the soundhole.

Fig. 6.84 An Ergo model guitar by Charles Fox showing a raised neck (image courtesy of Charles Fox, charlesfoxguitars.com)

6.2.1 Scale Length

Most acoustic guitars use one of a few common scale lengths. Of these, 25.4 in. (645.2 mm) and 25.5 in. (647.7 mm) may be the most common. The 25.4 in. scale is most closely associated with Martin and used on many of their models. The 25.5 in. scale is widely used, both acoustic and electric guitars, and is most closely associated with Fender. The obvious difference is simply the fret spacing and the size of the resulting instrument. However, there are some more subtle effects.

Tension in a stretched string is proportional to the square of length

$$T = 4f^2\rho L^2 \quad \text{or} \quad T\alpha L^2$$

where T is tension, f is frequency in Hz, ρ is mass per unit length, and L is length. Thus, longer strings need disproportionately higher tension as long as nothing else is changed. Longer strings also more closely behave like the ideal string approximation. The result of these factors and some others is that scale length can have a subtle, but important effect on the feel and tone of a guitar.

Some players state that a shorter scale can make a guitar's tone softer and warmer. This could be due to the dynamics of the strings themselves or to the structural differences in guitars with shorter scales. All else being equal, shorter strings have lower tension, which can make them easier to play.

Choices of scale lengths can also affect manufacturing decisions. Most manufacturers offer models with several different scale lengths and sometimes these lengths are related. Most Fender guitars have a 25.5 in. scale length, though a few, such as the Jaguar, use 24 in. (609.6 mm). Fender long scale basses use a 34 in. (863.6 mm) scale length and Squier (a Fender brand) currently offers a mini bass with a scale

Table 6.2 Fret locations based on 25.5 in. scale length

Fret	Dist from Nut	Dist from Bridge
−5	−8.538	*34.038*
−4	−6.628	32.128
−3	−4.825	30.325
−2	−3.123	*28.623*
−1	−1.516	27.016
0	0.000	*25.500*
1	1.431	*24.069*
2	2.782	22.718
3	4.057	21.443
4	5.261	20.239
5	6.397	19.103
6	7.469	18.031
7	8.481	17.019
8	9.436	16.064
9	10.338	15.162
10	11.189	14.311
11	11.992	13.508
12	12.750	12.750
13	13.466	12.034
14	14.141	11.359
15	14.779	10.721
16	15.380	10.120
17	15.948	9.552
18	16.484	9.016
19	16.990	8.510
20	17.468	8.032
21	17.919	7.581
22	18.344	7.156
23	18.746	6.754
24	19.125	6.375

length of 28.6 in. (726.4 mm). All these numbers are related in a way that seems unlikely to be a coincidence.

The expression for fret locations, measured from the nut is

$$L_i = L_0\left(1 - \frac{1}{r^i}\right)$$

where L_i is the position of fret i. For a guitar with 22 frets, i ranges from 0 to 22. The scale length is L_0 and r is $\sqrt[12]{2} \approx 1.05946$. Table 6.2 shows fret locations measured both from the nut and from the nominal location of the bridge (ignoring intonation), based on a 25.5 in. scale length. It may seem a little odd to let the fret number, i, be negative, but the equation doesn't mind. It is valid for both positive and negative values of i.

One can't help noticing that five different Fender scale lengths arise naturally, though approximately, from the math. In practice, this means that a gang saw with a single arbor for a 34 in. scale length could cut frets for 34 in., 28.6 in., 25.5 in., and 24 in. by simply shifting the fretboard blank with respect to the blades. Old timers at Fender acknowledge that such a saw did apparently exist.

Two obvious questions present themselves. The first is that the scale lengths are not exactly as referenced in lists of specifications. Fender basses are listed as having 34 in. scale lengths, not 34.038 in. scale lengths. Taking advertising literature as engineering literature is usually a mistake and so might it be here. Few players would know or care whether the listed scale length of their bass was off by 0.038 in. (1 mm). The other question is how the nut slot was cut. It is much wider than a fret slot, so perhaps a change in setup was needed.

While some manufacturers still use a gang saw to slot fretboards, others have moved to slotting using CNC tools. If the frets slots are cut with a CNC, it hardly matters whether there is any mathematical relationship between different scale lengths. They are all just implemented in software, so no one scale length is harder to cut than any other.

6.2.2 Neck Length

When designing guitars, one important design decision is the number of frets on the neck clear of the body. Classical guitars almost all have 12-fret necks. Steel string guitars usually have 14-fret necks, though a few models, often older designs, have 12-fret necks. The difference between them is apparent in two ways. The first is the obvious one in that 14 frets clear of the body offers the player better access to the higher frets. This was the original reason for Martin going to a 14-fret neck. Less obvious, though is the effect on the dynamics of the instrument.

If the body is not resized, the number of frets on the neck changes the location of the bridge, as shown in Fig. 6.85. The 14-fret neck pushes the bridge forward compared to a 12-fret neck, which changes how the vibrating strings excite vibrations in the soundboard. It's dangerous to generalize too much when dealing with structures as complicated as the braced soundboard of an acoustic guitar, but moving the bridge closer to the center of the vibrating part of soundboard tends to excite lower frequencies more effectively.

6.2.3 Headstock Shape

Headstock shape is one of the places that a designer has some aesthetic freedom. Headstock shape is often proprietary, serving as an identifying feature for that builder's instruments. In this sense, it's analogous to the unique shape of the

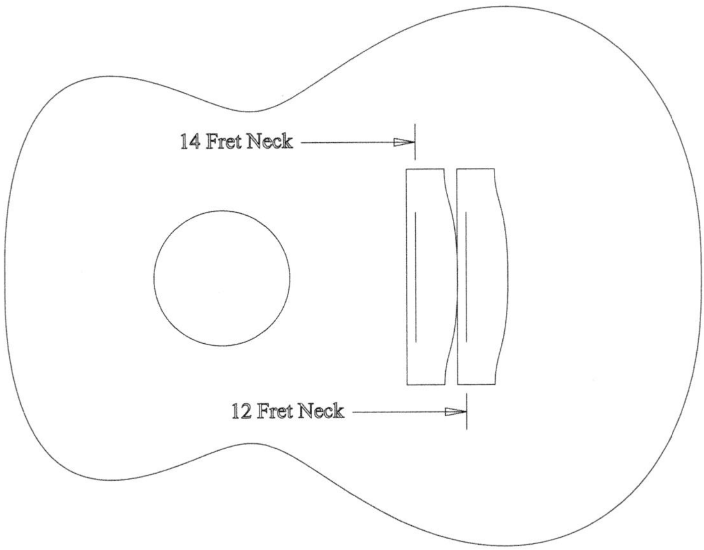

Fig. 6.85 Bridge placement with 12-fret and 14-fret neck

Coca-Cola bottle—anyone looking at the bottle knows right away what it is. Copying a headstock shape without permission is, as the very least, bad form. It may even constitute an infringement of a trademark, particularly if the builder sells the instrument with the copied headstock design. Fortunately, there is an infinite number of possible headstock designs and even the most generic will work just fine.

Perhaps the most familiar headstock shape is also the simplest. Since the 1800s, Martin has used a simple tapered design with a squared off end. A modern example on a Martin D-28 is shown in Fig. 6.86. This design is simple and classic. In combination with the Martin logo, it identifies the manufacturer to anyone who knows even a little about acoustic guitars.

A simple variation is shown in Fig. 6.87, on a guitar made by the author in a class taught by Charles Fox. This headstock has white and black veneers under the ebony headstock plate and a bevel on the end that highlights the contrast between them. Note also the inset washers on the head plate. This is a feature on most Fox guitars.

Figure 6.88 shows a generic headstock design, similar to the one above and reminiscent of the Martin design. This design shows tuners spaced 1.625 in on center. However, there is no standard tuner spacing for steel string guitars. The shortest common spacing is for 3-on-a-plate models with posts 1-3/8 in (34.9 mm) on center. 1-5/8 in (41.3 mm) is common for full sized guitars. The Stewart MacDonald plan of a Martin Spanish Style guitar from the 1840s shows a tuner spacing of 1.75 in (44.5 mm) on center, which is about the largest common spacing. Classical tuners do have a standard spacing, variously listed as either 1-3/8 in and 35 mm, though there is a slight difference between the two dimensions.

Fig. 6.86 Headstock of a Martin D-28 (image courtesy of Martin Guitars, martinguitars.com)

Fig. 6.87 A simple headstock with curved and beveled end

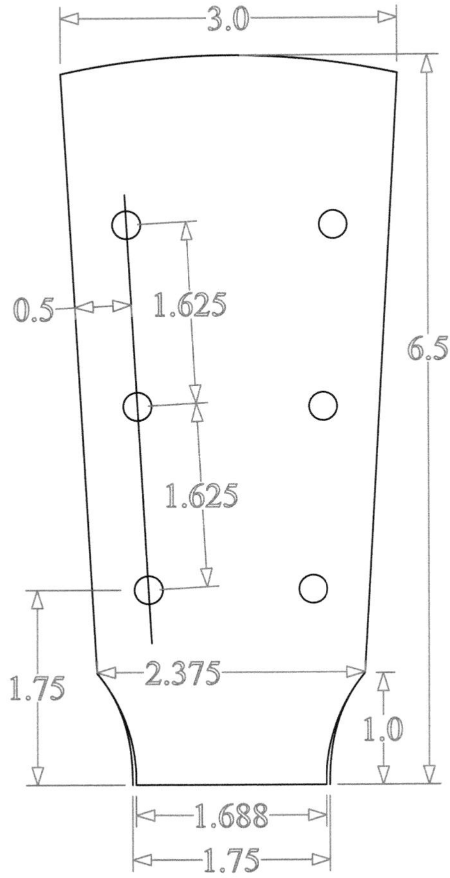

Fig. 6.88 Generic headstock design, dimensions in inches

The two most common nut widths in steel string guitars in the US are 1 11/16 in. (1.688 in. or 42.9 mm) and 1¾ in. (1.75 in. or 44.5 mm). Both are shown on this drawing. Thickness of headstocks on steel string acoustic guitars are generally between 0.575 in. and 0.625 in. (14.6–15.9 mm).

One concern when designing a solid headstock is that the inner strings can interfere with the string posts. The Martin headstock is carefully designed so that strings 3 and 4 come close, but don't touch. The string posts are not in a line parallel to the edge of the headstock. Rather, the middle two tuners are aligned with the lower two, as shown in Fig. 6.89.

There is no data in the literature to suggest that this contact would cause tuning problems, but it probably wouldn't help. A good solution is to reverse the taper of the headstock so that strings have a straighter path over the nut. Figure 6.90 shows the headstock of a Seagull guitar. The narrow headstock tapers toward the end so the strings pass nearly straight over the nut when viewed from the top.

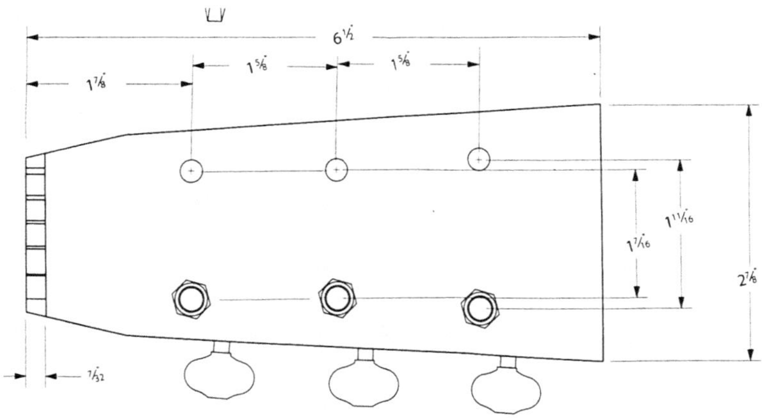

Fig. 6.89 Headstock of a 1933 Martin OM-28 (image courtesy of Stewart MacDonald, stewmac.com, plan by Don MacRostie)

Fig. 6.90 Headstock of a Seagull guitar (Wikimedia Commons, commons.wikimedia.org)

Some luthiers prefer that the strings pass as straight as possible over the nut. The reason often given is that it reduces tuning problems. While this may be correct, some of the most treasured acoustic guitars are Martins made before WWII and they all have the traditional headstock design.

A nice solution to this problem is found on flamenco guitars that use pegs for tuning. Since the peg doesn't care which way it turns, strings 1 and 6 are wound to the outside of the peg rather than the inside (Fig. 6.91). Note also that the distance between the pegs and the edge of the headstock is not constant. Since the knobs are on the back rather than the side of the headstock, the pegs don't need to be near the edge.

Fig. 6.91 Tuning pegs on a flamenco guitar (image courtesy of Paco Chorobo, chorobo.com)

Table 6.3 Names of the musical intervals

	1	2	3	4	5	6	7	8	9	10	11	12
1	Half Step	Whole										
2		Step	Minor	Major	Perfect	Aug.						
3			Third	Third	Fourth	Fourth	Perfect					
4						or	Fifth	Minor	Major	Minor	Major	Octave
5						Dim. Fifth		Sixth	Sixth	Seventh	Seventh	
6												
7												
8	Major	Minor	Major	Minor		Aug.						
9	Seventh	Seventh	Sixth	Sixth	Perfect	Fourth	Perfect					
10					Fifth	or	Fourth	Major				
11						Dim. Fifth		Third	Minor	Whole		
12									Third	Step	Half Step	

6.2.4 Fret Positions and Frequency Error

Small changes in pitch are more audible than small changes in amplitude. It seems obvious that the first step in building good tone into an acoustic guitar is to ensure that it makes the desired frequencies at each fret position. However, this is harder than it might first appear. Part of the problem is how musical notes are defined and part is how guitars are designed.

There is no such thing as being absolutely in tune, even theoretically, so designing a guitar is necessarily a succession of approximations and compromises. Western music arose from the Pythagorean ideal in which the frequencies of notes are related by ratios of small integers. The three most important ones are the octave (2:1), the perfect fifth (3:2), and the perfect fourth (4:3). The names of the musical intervals are shown in Table 6.3.

Table 6.4 Frequency ratios for the Pythagorean major scale

	C		D		E		F		G		A		B		C
Ratio	1		9/8		81/64		4/3		3/2		27/16		243/128		2
Interval		9/8		9/8		256/243		9/8		9/8		9/8		256/243	

For a Pythagorean major scale, defined in terms of increasing perfect fifths, the frequency ratios are shown in Table 6.4.

There are only two intervals, the whole step of 9/8 and the half step of 256/243. However, there is a big problem in that two half steps don't quite equal a whole step. That is 256/243 × 256/243 = 65,536/59049 ≈ 1.1099, but 9/8 = 1.125. The Pythagorean scale may be the starting point for western music, but it couldn't stop there because it is internally inconsistent and some combinations of notes sound out of tune. The solution was to slightly adjust the frequency ratios to make the system more workable. These adjustments are called temperament and the history of western music is, in part, the search for better forms of temperament. There are many tempering schemes and you may have heard of well tempering and just tempering. The one used by the guitar and other fretted instruments is equal temperament.

In equal temperament, the half step is defined as 1/12 of an octave. If we call r the frequency ratio of a half step, then $r = 2^{1/12} \approx 1.05946$, slightly different from the Pythagorean half step, where 256/243 = 1.053498. This way, 12 half steps naturally give an octave in which the frequency ratio is 2:1. A perfect fifth is 7 half steps and has a Pythagorean frequency ratio of 3/2 = 1.5. In equal temperament the frequency ratio is $r^7 = 1.4983$, so it's very close. However, it's not exact, so equal temperament is still an approximation.

A nice feature of equal temperament is that it allows frets to be straight and located in a regular pattern. Frets of modern instruments are almost always located assuming ideal strings and equal temperament, so that the distance from the nut to a fret is defined as

$$L_n = L_0\left(1 - \frac{1}{r^n}\right)$$

where L_0 is the scale length, n is the fret number, and $r = 2^{(1/12)} = 1.059463$. Distances are measured from the nut.

We tune guitars by the frequencies of open strings, assuming the fretted notes will then be in tune. However, the strings have to stretch a little when they are pressed against the fretboard. This slightly increases the tension, slightly increasing the pitch, even though the frets are placed using the equation above. If something isn't done, these increases in pitch can be large enough to be objectionable, so guitars, particularly those with steel strings, have slight modifications to reduce frequency errors. This is usually called compensation or intonation.

There are a surprising number of articles in the lutherie literature about temperament and correcting frequencies made by guitar strings. A few representative ones appear in the bibliography at the end of the chapter. Clearly, some bright people have spent time thinking about this problem.

By far, the most common compensation is applied to the saddle, as discussed earlier. However, there are more options available to guitar designers. From least to most effective, they are:

- None (unacceptable)
- Bridge compensation only (most common)
- Bridge compensation and nut offset
- Bridge and nut compensation
- Bridge and nut compensation along with offset frets

Refinements intended to reduce frequency errors in guitars are probably as old as the guitar itself. Frets on the earliest fretted instruments were tied on, so players could move them around to suit whatever temperament scheme they wanted. These gave way to fixed metal frets, though still with guts strings. Later, when steel strings were introduced, designers added more saddle compensation. Fixed frets, placed according to the ideal string equation, combined with saddle compensation is adequate for most players, but a few luthiers have introduced additional features to further reduce pitch errors.

Modern methods probably start with Bartolini and Bartolini in 1982, with Gilbert following in 1984. More recent articles by Greg Byers and Mike Doolin are also frequently referenced by luthiers.

Some simple modeling shows how compensation works for a plain steel string. Figure 6.92 shows calculated frequency errors for a plain steel B string at standard tuning with no modification to the instrument. It has a diameter of 0.016 in. (0.406 mm), a length of 25.5 in. (647.7 mm), and a target frequency of 246.94 Hz. Frequency ratio is the actual frequency the string makes divided by the one defined by equal tempering. Ideally, the frequency ratio should be 1 at every fret, meaning that the fundamental frequency of the string each fret is exactly matches the ideal. Note that errors are much smaller for nylon strings since the elastic modulus of nylon is about 30× less than that of steel.

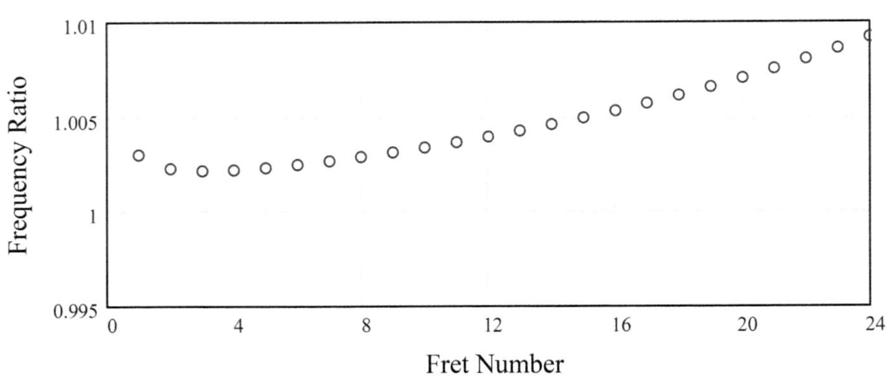

Fig. 6.92 Calculated frequency error for B string with no compensation

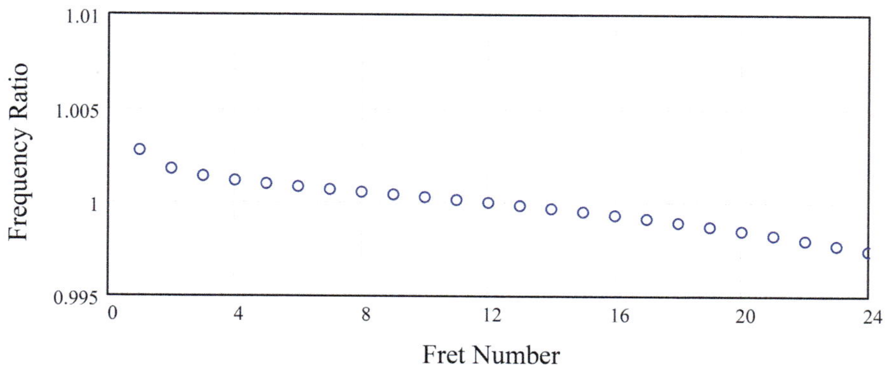

Fig. 6.93 Calculated frequency error for a B string with saddle compensation

These errors are large enough to be audible, something luthiers quickly realized when steel strings became available. Their empirical solution was to move the saddle back slightly to lower the pitch of the strings when fretted. The effect is magnified for shorter strings, so saddle intonation is most effective at the higher frets, right where it is most needed. Mathematically moving the saddle back by 0.099 in. (2.52 mm) greatly reduces the predicted frequency errors, as shown in Fig. 6.93.

While calculated using some assumptions, this is fairly close to what guitarists experience. The predicted frequencies are much closer to the ideal ones. For reference, the frequencies of the notes in the human hearing range are shown in Table 6.5. Frequencies of the open strings, in standard tuning, are highlighted.

The nut can also be intonated by moving it slightly closer to the saddle. The model predicts that saddle compensation tends to rotate the error curve, while nut compensation tends to shift it vertically (Fig. 6.94).

Figure 6.95 shows the calculated frequency error for the same G string with 0.11 in. (2.8 mm) compensation at the saddle and 0.05 in. (1.27 mm) at the nut. The nut compensation moves the nut closer to the saddle. In this case, the predicted frequency error is significantly decreased everywhere but the first two frets.

These calculations will give slightly different results for each of the six strings because they are different diameters, have different target frequencies, and may be wound to increase mass. An approximate approach is to ignore differences between the strings and simply shift the nut slightly closer to the saddle without any compensation for the individual strings. This means just trimming the fretboard a little short at the nut or moving the nut slot. This refinement sometimes appears both in factory-made instruments and those made by individual luthiers. Exact offsets are understandably proprietary, but seem to be in the range of 0.030–0.060 in. (0.76–1.5 mm).

Most methods for reducing frequency errors in guitars focus on compensation at the nut and saddle, usually string by string. These have the advantage of not requiring large modifications and most work well enough to make noticeable improvements. While all production acoustic guitars have saddle compensation and many

Table 6.5 Frequencies of notes in the equal tempered scale (Hz)

E_0	20.602	E_3	164.81	E_6	1318.5	E_9	10548
F_0	21.827	F_3	174.61	F_6	1396.9	F_9	11175
	23.125		185.00		1480.0		11840
G_0	24.500	G_3	196.00	G_6	1568.0	G_9	12544
	25.957		207.65		1661.2		13290
A_0	27.500	A_3	220.00	A_6	1760.0	A_9	14080
	29.135		233.08		1864.7		14917
B_0	30.868	B_3	246.94	B_6	1975.5	B_9	15804
C_1	32.703	C_4	261.63	C_7	2093.0	C_{10}	16744
	34.648		277.18		2217.5		17740
D_1	36.708	D_4	293.66	D_7	2349.3	D_{10}	18795
	38.891		311.13		2489.0		19912
E_1	41.203	E_4	329.63	E_7	2637.0	E_{10}	21096
F_1	43.654	F_4	349.23	F_7	2793.8		
	46.249		369.99		2960.0		
G_1	48.999	G_4	392.00	G_7	3136.0		
	51.913		415.30		3322.4		
A_1	55.000	A_4	440.00	A_7	3520.0		
	58.270		466.16		3729.3		
B_1	61.735	B_4	493.88	B_7	3951.1		
C_2	65.406	C_5	523.25	C_8	4186.0		
	69.296		554.37		4434.9		
D_2	73.416	D_5	587.33	D_8	4698.6		
	77.782		622.25		4978.0		
E_2	82.407	E_5	659.26	E_8	5274.0		
F_2	87.307	F_5	698.46	F_8	5587.7		
	92.499		739.99		5919.9		
G_2	97.999	G_5	783.99	G_8	6271.9		
	103.83		830.61		6644.9		
A_2	110.00	A_5	880.00	A_8	7040.0		
	116.54		932.33		7458.6		
B_2	123.47	B_5	987.8	B_8	7902.1		
C_3	130.81	C_6	1046.5	C_9	8372.0		
	138.59		1108.7		8869.8		
D_3	146.83	D_6	1174.7	D_9	9397.3		
	155.56		1244.5		9956.1		

slightly offset the nut, production guitars with compensated nuts are rare indeed. For now, nut compensation is the realm of individual luthiers.

A method for nut and saddle compensation, developed by Trevor Gore, uses measured frequency errors combined with calculations to optimize the nut and

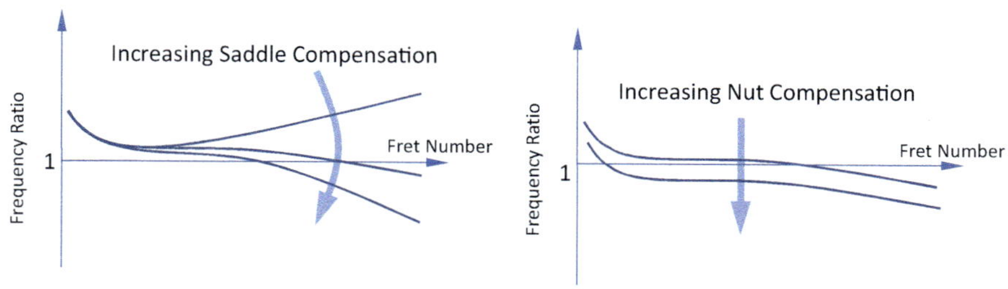

Fig. 6.94 Effect of nut and saddle compensation

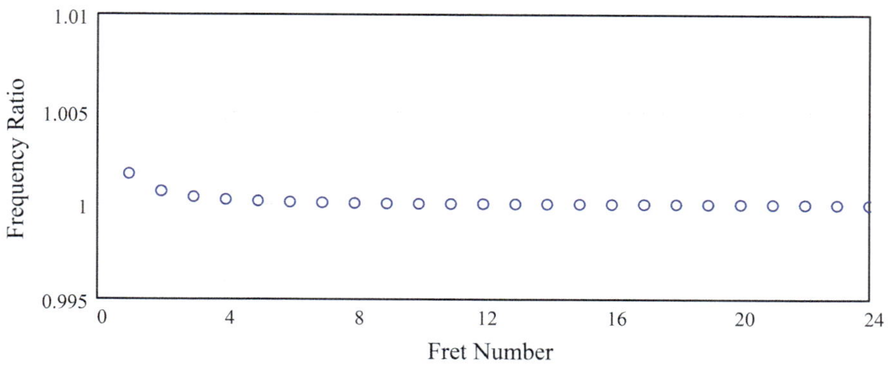

Fig. 6.95 Calculated frequency error for a B string with nut and saddle compensation

saddle offsets for each string. It is described in detail in his book, co-authored with Gerard Gilet, Contemporary Acoustic Guitar Design and Build, Vol 1. His experience with his own guitars and with modifying production instruments has shown the effectiveness of this method. Figure 6.96 shows the compensated nut on a guitar made by Gore. Note also the headstock design that gives the strings a straight path over the nut when viewed from the top.

In the absence of measurements or a better math model, Gore suggests applying half the nominal saddle compensation to the nut and applying the rest to the saddle. While not as effective as calculated offsets, it should still be an improvement over simply offsetting the nut.

Even with compensation at the nut and saddle, there can be noticeable frequency errors at the first two frets. Slightly shifting the two first frets toward the nut eliminates these predicted frequency errors, as shown in Fig. 6.95. Figure 6.97 shows the predicted effect on the B string of moving the first fret 0.040 in. (1.02 mm) closer to the nut and the second fret 0.015 in. (0.38 mm) closer to the nut.

The results of these calculations depend on several physical parameters, including string diameter, string mass, string height, and scale length. In order for the calculated results to be useful, we need some practical guidelines. For a steel string guitar with scale length near 25.5 in. (647.7 mm), the following modifications are reasonable:

Fig. 6.96 Intonated nut on a guitar by Trevor Gore (image courtesy of Trevor Gore, goreguitars.com.au)

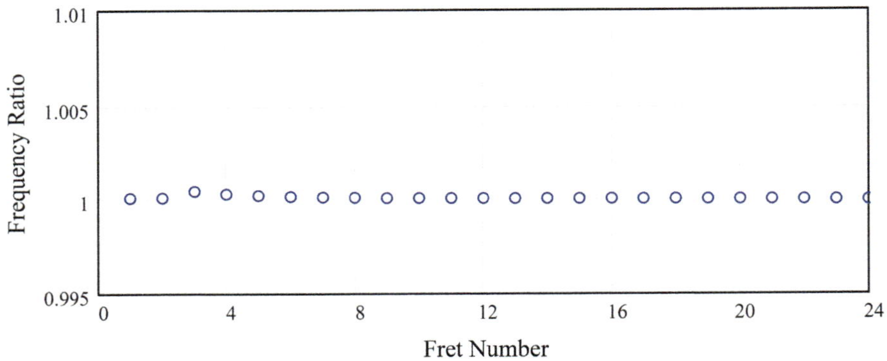

Fig. 6.97 Calculated frequency error for a B string with nut and saddle compensation and shifted frets

- Move nut 0.050 in. (1.3 mm) closer to the saddle.
- Move the first fret 0.050 in. (1.3 mm) closer to the nut.
- Mover the second fret 0.020 in. (0.5 mm) closer to the nut.
- Compensate the nut according to the diameter of the string, or the core on a wound string.
  - String 1 (high E) gets 0.050 in. (1.3 mm) compensation so its length is unchanged.
  - String 6 (low E) gets no compensation, so its length, before saddle compensation, is reduced by the nut offset.
  - Compensation for strings 2–5 set by measurements at fret 12 using a tuner.

    Reducing nut compensation (shifting toward nominal nut position) sharpens string.
    String 2 compensation a little less than half of nut offset.
    String 3 a little less than nut offset.

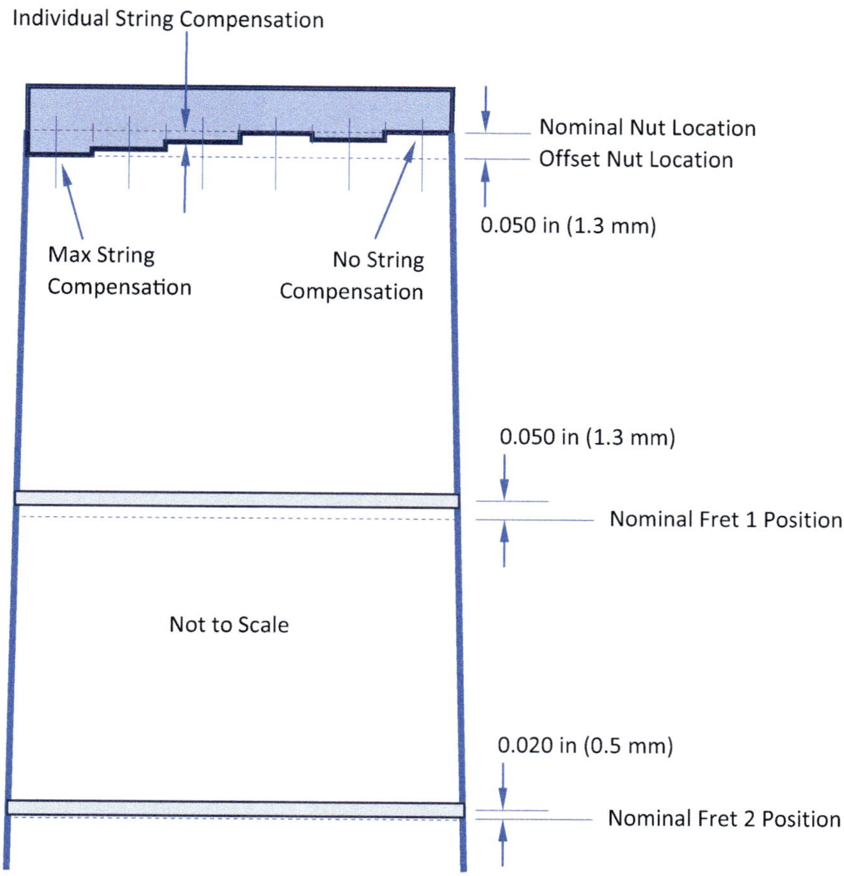

Fig. 6.98 Nut compensation and fret offset for a steel string guitar

More specific instructions require more knowledge about the size of the strings, their height above the frets, and scale length. A few guidelines:

- Shorter scale length requires more compensation.
- Shorter scale length requires larger shift of first two frets.
- Higher action requires more compensation.
- Higher action requires larger shift of first two frets.

This approximate approach has been used on a number of electric and acoustic guitars. Measurements have shown lower frequency errors and subjective evaluations have shown that players generally prefer instruments with this compensation. Figure 6.98 shows the modifications. The thinner the core of the string, the less compensation it needs. String 1 (high E) needs almost none, so the compensation for the that string is about equal to the offset of the nut. Note that there is also little compensation at the saddle for this string.

Fig. 6.99 Compensated nut on an acoustic guitar

Figure 6.99 shows the nut on an acoustic guitar compensated using this method.

Greg Byers developed a method of nut and saddle compensation, based on both calculation and measurements, that he applied to classical guitars. His results are similar to those proposed for steel string guitars. Results for a classical guitar with a scale length of 650 mm (25.6 in.) are shown in Fig. 6.100. Nylon classical strings differ from steel strings in few ways that affect compensation. The wound strings have stranded cores that have smaller effective diameters than it may appear. Also, the solid nylon treble strings have very large diameters compared with steel strings. Solid G strings typically have a diameter around 0.035–0.040 in. (0.89–1.0 mm). It is not surprising that the offset roughly follows core diameter, with the solid G string needing the most compensation.

A broader approach to compensation, developed by Gary Magliari and implemented by Don MacRostie, calculates fret locations for each string to minimize frequency errors. In order to allow straight frets, a minimum error position is calculated for each fret. Don MacRostie, of Red Diamond Mandolins and formerly of Stewart MacDonald (retired), has implemented it on mandolins with very good results. Maglilari has since refined his approach based on the idea of tension length, the distance from the tuner to the tailpiece. Tension length is longer than the vibrating length of the string.

Finally, we should note that some builders have gone one step further and gone to segmented frets or frets that aren't straight. A few luthiers have made fretboards in which each fret is in six segments that don't always line up. One manufacturer, True Temperament, has developed a fret system around cast frets that are not straight

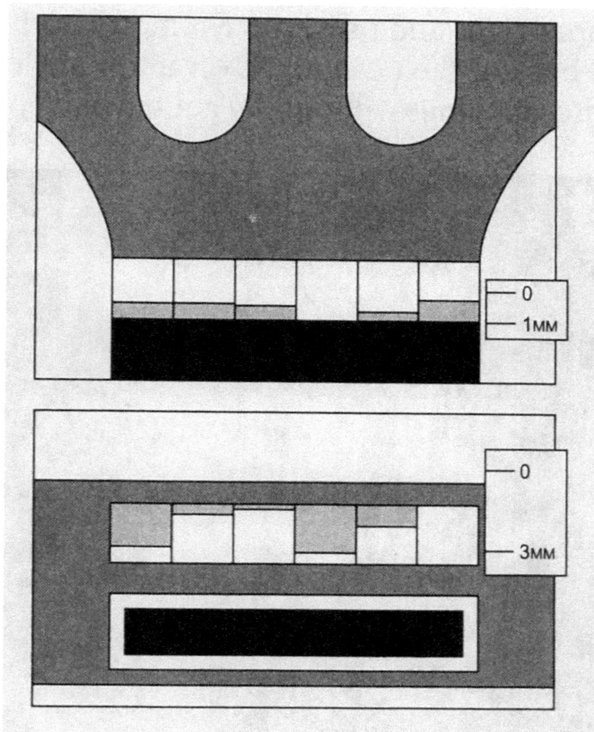

Fig. 6.100 Nut and saddle compensation for a classical guitar (image courtesy of Guild of American Luthiers, luth.org, image by Greg Byers)

and fit into matching pockets cut into the fretboard using a CNC router. The attraction of straight frets has, so far, outweighed the benefits of systems requiring segmented or bent ones.

6.2.5 *Tuning Machines*

Tuning machines, sometimes also called machine heads, are almost always just worm gears. Whatever their refinements, they are based on a very old idea. The worm gear or "endless screw" appears to have originated in ancient Greece and may have been invented by Archimedes. They appear in many types of machinery and are about as common as machine elements get. Figure 6.101 shows a worm gear that operates part of a canal lock.

The pitch of the string is a function of tension

$$f = \frac{1}{2L}\sqrt{\frac{T}{\rho}}$$

Fig. 6.101 A worm gear operating part of a canal lock (Wikimedia Commons, commons.wikimedia.com)

where T is tension, L is the length of the string, and ρ is the mass per unit length of the string. The tension is, in turn, a function of string length, diameter, and elastic modulus, E. Elastic modulus is essentially the stiffness of the material. The change in tension is

$$\Delta T = \frac{\Delta LAE}{L}$$

where ΔL is the change in length, E is the elastic modulus, A is the cross-sectional area, and L is the scale length. Note that A is the area of the load carrying part of the string only. For a wound string, the load carrying area is the core only.

Each turn of the post increases the length of string by the circumference of the post or roller around which the string is wound. Classical tuners generally have 10 mm (approx. 3/8″) diameter rollers and steel string tuners usually have posts of either 6 mm (0.236 in.) or ¼ in. (6.35 mm). The gear ratios of classical and steel string tuners are about the same, with ratios usually in the range of 14:1 to 18:1. Higher ratios should give more precise tuning since they give more precise change in ΔL and thus more precise change in tension. At this writing, Kluson offers a tuner with 19:1 ratio. Gotoh and Waverly offer ones with 21:1 ratio.

Note that the change in tension for a given change in length of the string is smaller for longer scale lengths (because L is in the denominator). This means that pitch on guitars with longer scale lengths changes less due to a given rotation of the tuner post than those with shorter scale lengths. In practice, this means that short scale length instruments should have tuners with higher tuning ratios. For example, the 21:1 ratio tuners would be well suited to a short scale length instrument.

Figure 6.102 shows a Sta-Tite tuner by Grover, a popular tuner for acoustic guitars. The open worm gear is clearly visible. While it's much smaller than the one working the canal lock, it's clearly the same machine.

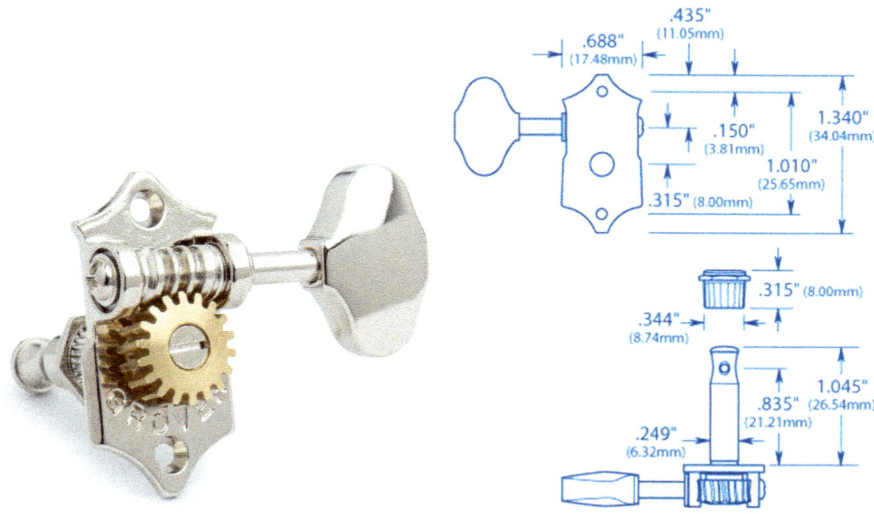

Fig. 6.102 A Sta-Tite 18:1 tuner by Grover (images courtesy of Stewart MacDonald, stewmac.com)

Most tuning guitar machines have sealed gearboxes that enclose lubricated gears. Rotomatics, also by Grover, are a popular enclosed tuner (Fig. 6.103).

When choosing tuners, there are only a few decisions for the designer:

- Open or closed
- Nylon or Steel Strings
- Solid or slotted headstock
- Gear ratio

The choice of open or closed gear often revolves around aesthetics and weight. Open gear tuners are a nod to older designs. However, they are also lighter. The reduced weight slightly affects balance and may even affect the dynamics of the neck. However, most tuners on steel string guitars have enclosed, permanently lubricated gears.

A sort of middle ground between open gear and closed gear machines are those fitted with stamped covers. These are mechanically similar to open gear machines, but with covers to protect the gears. While produced by a number of different manufacturers, this style is associated with Kluson. Fig. 6.104 shows a Kluson Deluxe tuning machine with a stamped cover.

A variation on the open gear tuners and those with stamped covers are those mounted three to a plate, as shown in Fig. 6.105. These tuners are Golden Age brand, but closely follow the Kluson pattern. There is a single plate for each side of the headstock that mounts three tuners as one unit.

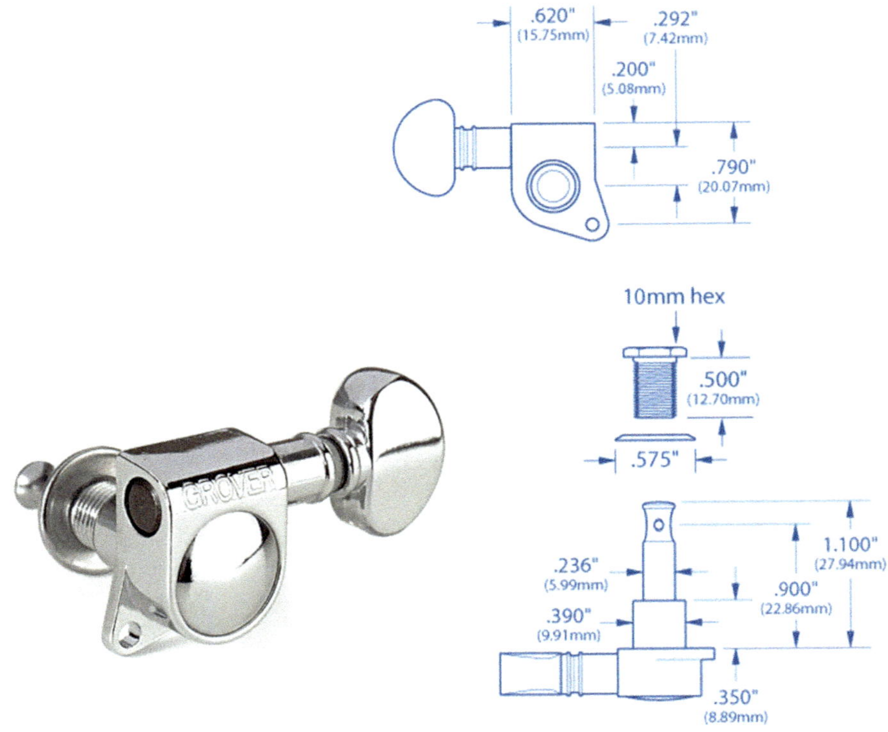

Fig. 6.103 A Grover Mid-Size Rotomatic tuner (images courtesy of Stewart MacDonald, stewmac.com)

Fig. 6.104 Kluson tuning machine with stamped cover (image courtesy of Stewart MacDonald, stewmac.com)

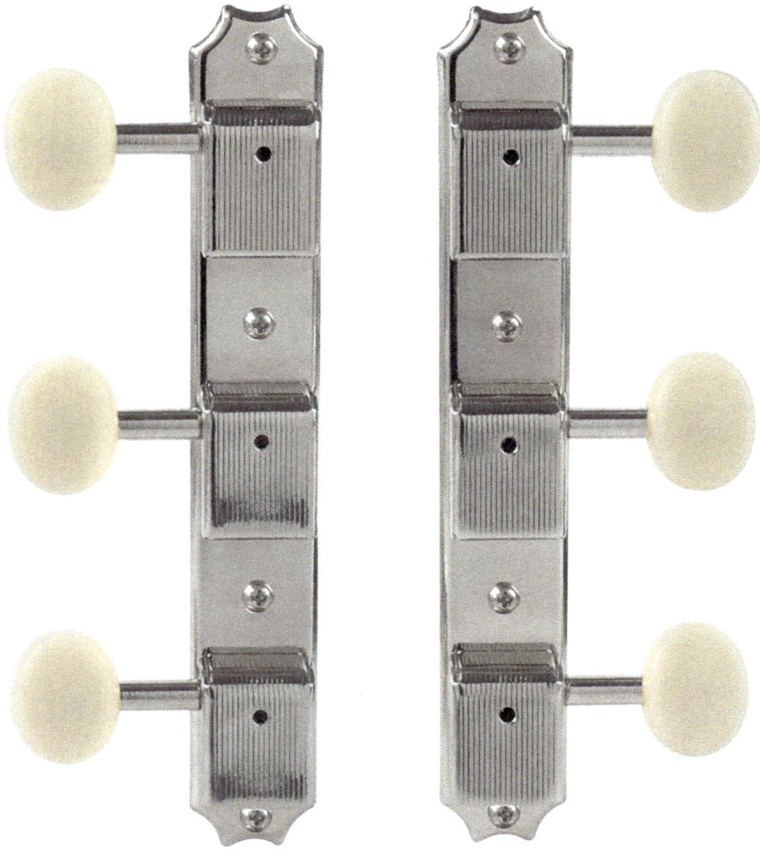

Fig. 6.105 Golden Age Three on a plate tuners (image courtesy of Stewart MacDonald, stewmac.com)

Older steel string guitars and almost all classical guitars use a slotted headstock, in which the gears are mounted on the side of the headstock rather than on the back. The posts are accessed through slots in the headstock. This design is now rare on steel string guitars, being used mostly on instruments meant to be reminiscent of older designs. Figure 6.106 shows a Taylor 322e with a slotted headstock and a 12-fret neck.

Tuners with plain posts rely on friction to keep the strings from slipping around the posts and going slightly flat. Some tuners have locking mechanisms to fix the end of the string in the hole in the post (Fig. 6.107). While largely eliminating string slippage, they are heavier and more expensive than conventional tuners.

Fig. 6.106 A Taylor 322e guitar with a slotted headstock (image courtesy of Taylor Guitars, taylorguitars.com)

Fig. 6.107 Grover Roto-Grip Locking Tuner (image courtesy of Sweetwater Sound, sweetwater.com)

A few tuning machines use something other than worm gears. An interesting design for solid headstocks is a gearless tuner by Steinberger (Fig. 6.108). This tuner uses an internal threaded post to pull the string downward through the barrel. The fine threads in the barrel allow for precise tuning.

Traditional flamenco guitars use wooden pegs for tuning. Flamenco guitars were originally designed to be very inexpensive and wooden pegs were the cheapest option. However, flamenco instruments are now often built for players who want a fine instrument. To keep the look of wooden pegs while offering the refinement of geared tuners, a few manufacturers offer pegs with internal gearing. Figure 6.109 shows geared tuning pegs on a flamenco guitar by Paco Chorobo. There are only a few manufacturers of geared tuning pegs and the gear ratio is lower than that for other types of guitar tuners. Those made by Wittner have an 8.5:1 ratio. However, the smaller diameter of the peg effectively increases the ratio, compared to classical guitar tuners with 10 mm diameter rollers.

Perhaps the most important characteristic of tuning machines is the gear ratio. The gear ratio defines how many times the knob must be turned to turn the post

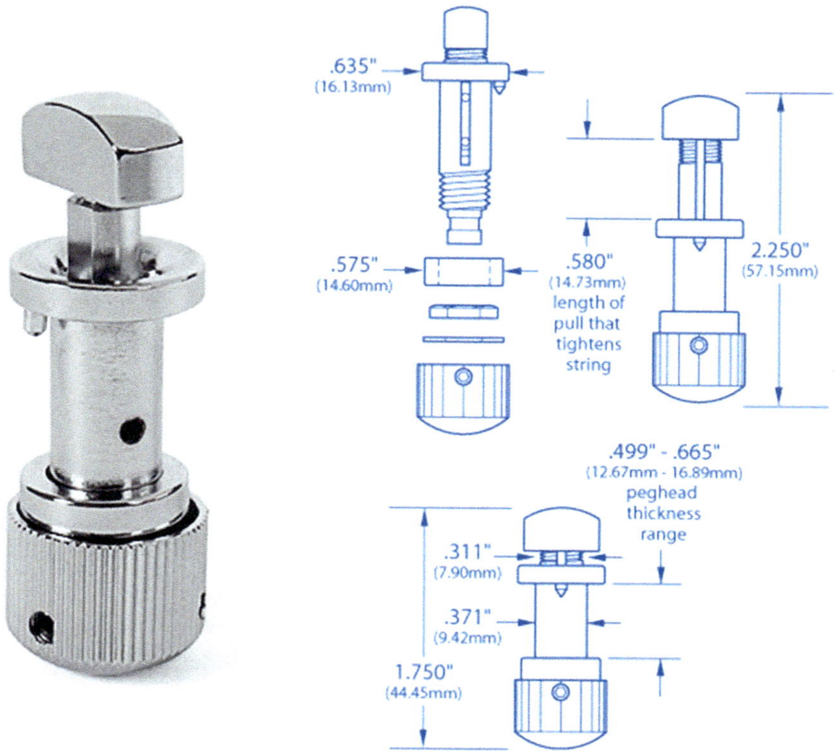

Fig. 6.108 Steinberger gearless tuner (images courtesy of Stewart MacDonald, stewmac.com)

Fig. 6.109 Geared tuning pegs on a flamenco guitar (image courtesy of Paco Chorobo, Chorobo.com)

once. It's important to understand a subtle mathematical feature of tuning machines. They are designed to stretch the string in order to change the tension. They directly change the length of the string and the change in tension is an indirect effect; it basically comes along for the ride. The change in tension is what changes the pitch of

Fig. 6.110 Gear Ratios of GraphTech acoustic tuners (image courtesy of GraphTech, graphtech.com)

Fig. 6.111 A set of open back acoustic guitar tuners with different gear ratios (image courtesy of GraphTech, graphtech.com)

the string. This may seem like a picky distinction, but it is central to how guitars are tuned.

Basic physics dictates that each of the six strings on a guitar requires a different rotation of the knob for a given change in pitch. The heavier, low pitch strings require fewer turns than lighter ones. A clever solution, by GraphTech in their Ratio tuners, is to give each tuner a different gear ratio. Figure 6.110 shows the gears, from treble E on the left to Bass E on the right.

Figure 6.111 shows a set of open back Ratio tuners. The differing gear ratios are clearly visible.

Few things in the guitar world are standardized and this is perhaps most apparent in how tuners are mounted. Tuners have small screws or pins to keep them from rotating under string tension and several mutually incompatible geometries are commonly used. Closed back Ratio tuners come with a selection of light aluminum

Fig. 6.112 A Fender CF-140S retrofitted with Ratio tuners

adapter plates that allow them to be retrofitted to most guitars without modification. It's worth noting that four different sets of plates are supplied. Figure 6.112 shows a set of Ratio retrofitted to a Fender CF-140S with the aid of the correct adapter plates.

Finally, there are a few small manufacturers who make very-high-quality tuning machines, often building to order. One of these is Rodgers Tuning Machines, who offer a range of tuners for both classical and steel string guitars as well as ukuleles. They offer a wide range of designs, with many options, both mechanical and aesthetic. Figure 6.113 shows a set of Rodgers tuners for a steel string guitar.

6.3 Aesthetics

Few parts of guitar design are more personal than the aesthetic elements of the instrument. Some players prefer a simple, classic look while others are the opposite, drawn to rare woods and elaborate decorative elements. Fig. 6.114 shows a Taylor PS14ce, a guitar in their Presentation Series. The top is made from sinker redwood (from a log submerged in a California riverbed) while the back and sides are from

Fig. 6.113 A set of Rodgers steel string tuning machines (image courtesy of Rodgers Tuning Machines, rodgers-tuning-machines.com—permission needed)

Fig. 6.114 A Taylor PS14cd (image courtesy of Taylor Guitars, taylorguitars.com)

Tasmanian blackwood. There are elaborate inlays on the neck and bridge, along with inlaid rosette and binding. This is a striking instrument.

Taylor is well known for their early introduction of computer-controlled milling machines and it's likely that the inlays in the instrument shown here were at least partly made using a CNC mill. This makes sense for a company what will make many such inlaid instruments and can design the inlays on CAD and produce them with the assistance of CNC milling.

Such elaborate inlays are found less often on guitars made by individual luthiers as they can be spectacular time sinks. It's more common for luthiers to limit inlays to a few well-executed elements. Figure 6.115 shows a nice inlay on a Turnstone guitar by Rosie Heydenrych. Note also the silver outlined side dots.

Fig. 6.115 An inlay at the 12th fret of a Turnstone guitar (image courtesy of Rosie Heydenrych, turnstoneguitar.co.uk)

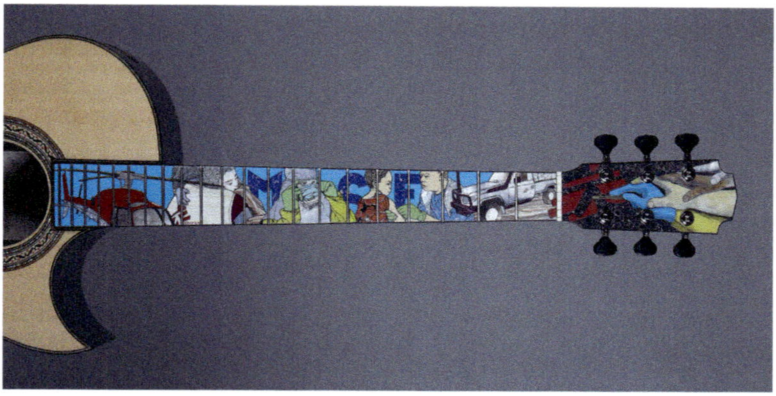

Fig. 6.116 An elaborately inlaid fretboard and headstock (image courtesy of Grit Laskin, williamlaskin.com)

A few luthiers are known for extremely elaborate inlays that require patience, creativity, and the highest levels of skill. One of most well-known among this small group is Grit Laskin. Figure 6.116 shows one of his guitars with inlaid fretboard and headstock as a tribute to Médicins Sans Frontiéres (Doctors without Borders). The instrument was sold to support MSF, an organization that Laskin greatly admires.

It's worth noting that inlay does nothing for the tone of the instrument. It's strictly decorative. However, the finest guitars can be artistic statements as much as they are musical ones. As artists, it's important that luthiers express their aesthetic visions as they bring their instruments into being.

6.4 Examples of Design Refinement

While some luthiers focus on traditional designs, others seek constant refinement. Refinements are often subtle and they must contribute to the whole. A refined design cannot simply be a collection of features. Rather, a coherent, sophisticated design needs to incorporate the features so they work together in a way that creates a fine instrument.

The best luthiers pay attention to details that make their instruments more comfortable to play, features that are sometimes absent from production instruments. Figure 6.117 shows an acoustic guitar by Jay Lichty that includes a wedge body and a radiused arm rest. Note that the body is shallower on the left side of this picture than on the right. This feature, originally developed by Linda Manzer, is intended to allow the instrument to fit more naturally under the arm of the player. Note also the larger radius on the left side of the body—the upper corner of the lower bout. This variation on the arm bevel, which was originated by Grit Laskin, keeps the edge of the instrument from digging into the player's arm.

An example of a design that represents many years of experience and the accumulated wisdom of many experiments is the Ergo model guitar by Charles Fox (Fig. 6.118). Top luthiers develop their designs as they develop a deeper understanding of their craft and Charles is certainly one of those. This guitar incorporates a number of design refinements, some visible to the player and some not, and doing so in a way that they complement one another.

Finally, Fig. 6.119 shows a Turnstone Model TS, by English luthier Rosie Heydenrych. It incorporates offset fret markers, a partial cutaway, a live back and a small arm bevel.

Fig. 6.117 Wedge body in a guitar by Jay Lichty (image courtesy of Corrie Woods, lichtyguitars.com)

Fig. 6.118 An Ergo guitar by Charles Fox (image Courtesy of Charles Fox, charlesfoxguitars.com)

Fig. 6.119 Turnstone Model TS Guitar (image courtesy of Rosie Heydenrych, turnstoneguitar.co.uk)

6.5 Durability

Guitars lead widely varied lives. Some are pampered—handled gently, maintained carefully, and stored under ideal conditions. Others are almost literally played to pieces—played hard, given little maintenance, and stored in whatever conditions are convenient. Guitars should last, even under heavy, but reasonable use.

We need to distinguish between heavy use and willful, destructive abuse. Pete Townshend smashing a guitar on stage is destructive, willful abuse. No guitar should be expected to hold up to that. Rather, acoustic guitar designers try to produce guitars that are good instruments and can hold up to long use, even in conditions that aren't ideal.

Guitars should be played. They are safe when stored in a case or hanging on the wall, but that's not what they are for. Unlike violins, guitars aren't generally designed to be disassembled for maintenance (classical guitars assembled with hot hide glue are an exception). That means they wear out eventually. Repairs can extend the working life of acoustic guitars, but that life is still finite, seldom longer than the working life of the player.

On the far end of this spectrum is Trigger, Willie Nelson's famous 1969 Martin N-20 classical guitar (Fig. 6.120). It undergoes regular maintenance by Mark Erlewine and is cared for on tour by Nelson's guitar tech. The N-20 was not commercially successful, but Nelson's love of this otherwise unremarkable instrument is the stuff of guitar legend. As a classical guitar, it was designed to be plucked with

Fig. 6.120 Willie Nelson playing Trigger at an outdoor concert in 2009 (Wikimedia Commons, commons.wikimedia.com, uploaded by Bob Tilden)

fingertips and it originally had no pickup. It was modified to add a piezo electric bridge pickup and Nelson played it with a pick right from the beginning. Since there is no pickguard, he quickly began to wear away the finish and then wood of the top between the soundhole and the bridge. He eventually wore a hole through the top, exposing the bracing.

By any reasonable definition, this guitar is worn out. Indeed, people have tried to convince Nelson to replace it with a faithful copy, right down to the pickup. Alas, he won't have it. The replacement guitar wouldn't be Trigger and that's what matters to him. It seems that, in Trigger, he finds his musical voice. Only careful maintenance is keeping this instrument working.

My own opinion is that a guitar never played is just as tragic as a book that is never read or a ship that never puts to sea. Keeping a guitar safe, but mute rather misses the point and might even be an insult to the builder. Take care of your guitar, but play it. Play it to pieces. When it can no longer be fixed, hang it on the wall for display and go get another guitar.

Wear on acoustic guitars is often of only a few types:

- Worn or damaged finish
- Permanent deformation that distorts the soundboard or changes neck angle
- Cracks
- Bridge lifting
- Worn frets and fretboards

The designer has essentially two options when addressing these common failures. The first is to make them easier to repair and the second is to prevent them from happening in the first place. Modern designs have evolved under disparate pressures, such as improved tone and ease of manufacture. Durability is another of these pressures, since no one wants to develop the reputation of making unacceptably fragile guitars.

As with many things in engineering, the various constraints tend to work against each other. For example, it's easy to make an instrument durable by making it much heavier and stronger. Of course, the resulting tone is likely to be quite bad. The result is that designers succeed, in part, by balancing conflicting requirements. Let's examine the types of failures individually.

6.5.1 Finish

Few elements of the acoustic guitar have changed more than finish. The early guitar finishes were shellac, lacquer, and varnish. They are durable enough to protect the instrument without having to be so thick as to degrade the tone. They are easy enough to apply and, perhaps more important, they are easy to repair. More modern catalyzed finishes are harder and more durable. However, they are more difficult to apply and to repair.

Fig. 6.121 A 1936 Martin OO-18 showing typical wear (image courtesy of Mike and Mike's Guitar Bar, mmguitarbar.com)

Worn finish is common on old guitars and ones that have been played a lot. It's not surprising that the finish wears most where is it in contact with the player. Sometimes the finish is worn through, though seldom to the point of significantly abrading the wood. Figure 6.121 shows a prewar Martin OO-18 with some discoloration where the players right arm contacts the lower bout and finish worn off the upper bout on either side of the neck. Apart from this, it has the usual dents and scratches one would expect on an old guitar that has been played. This one didn't live in a glass case; it has worked for a living. It is interesting to note that the wear on the upper bout is above the pick guard, which is intended to prevent just this kind of thing. Apparently, the player preferred to pick farther up the strings from the bridge than is common.

Repairing worn finish can be as simple as just applying some new finish where the old has been worn away. This is particularly easy with shellac and lacquer. They are evaporative, so new finish simply melts in as the solvent softens existing finish. Other types of finish may require some surface preparation. The nature of finish repairs depends on the instrument. Large finish repairs that are acceptable for a working instrument might not be for one with some historical importance.

It can be difficult to perfectly blend finish repairs so that they are invisible. For example, it's not uncommon for edges of repairs on hard catalyzed finishes to show under close inspection. How big a problem this is depends on the needs of the

Fig. 6.122 Paco Chorobo applying a clear tap plate to a flamenco guitar (image courtesy of Paco Chorobo, chorobo.com)

owner. For a working guitar on which finish repairs are just a matter of protecting the wood, small imperfections might be acceptable. Small chips in hard, production finishes can often be "drop filled" with superglue, then carefully sanded and buffed. For valuable instruments, repairs might need to be nearly invisible, requiring more time and carefully considered methods.

Good designers should anticipate the most common types of failure, like worn finish. The most obvious design feature on acoustic guitars is pick guards, like the one in Fig. 6.121. This is nothing more than a thin sheet of plastic glued to the top where a pick is most likely to hit it. Willie Nelson's Trigger (Fig. 6.120) is the perfect example of what can happen to a guitar without a pickguard when the player uses one anyway. Pickguards were originally celluloid, a thermoplastic formed by mixing nitrocellulose and camphor. Moderns pickguards are often made of plastics like PVC.

A variation on this idea are the thin, often clear, plastic tap plates on flamenco guitars. Flamenco guitars are almost never played with picks, but the flamenco players often tap the top with their fingernails as part of the music, which could damage the soft wood tops. Figure 6.122 shows Paco Chorobo applying a tap plate to a flamenco guitar. The tap plate is clear and just visible in the center of the body.

The sides and back of an acoustic guitar contribute much less to tone than does the top. Thus, it's not a problem to put thicker finish on the back and sides. The problem is that the back and sides don't usually see the most wear.

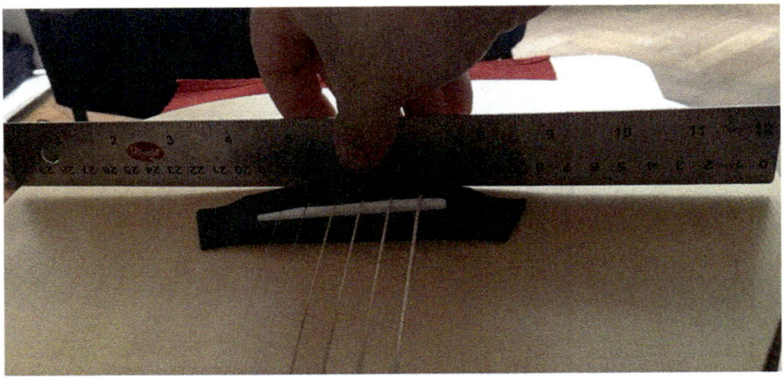

Fig. 6.123 Curvature of soundboard behind bridge (retrieved from Acoustic Guitar Forum, acsouticguitarforum.com, image posted by user pflopez)

6.5.2 *Permanent Deformation*

Wood creeps under load, permanently changing shape over time, even though it's not at failure load. Creep in wood depends on the loading direction. Wood tends to creep more in the tangential direction than in the radial direction. Creep in the longitudinal direction is generally low. All this means that acoustic guitars should be expected to slightly change shape over long periods.

It's common for flattop guitars to develop a "belly" or depression between the bridge and soundhole as a result of the string tension applying a torque to the bridge. The soundboard also bulges behind the bridge (Fig. 6.123). Some of this is elastic response to string tension, but it can become permanent over time. This shouldn't be surprising since the top plate is thin and bending stiffness is a strong function of thickness. Also, the soundhole is centered under the strings, right where the load path would otherwise lie.

Designing against this type of failure is difficult. A structure heavy enough not to deform significantly over decades of steady loading might be so heavy as to hurt the tone of the instrument. Moving the soundhole off the centerline of the instrument would let the designer place bracing along the load path. It is also possible to stiffen the top where the curvature from the deformation is the highest, either by leaving the top thicker or modifying the bracing. Of course, another possibility is to make the top from a material less susceptible to creep.

At some point, a designer needs to ask how long the guitar should last. With careful design and diligent maintenance, a lightly built, but responsive guitar might have a working life of 20 years or so. A heavier structure, if well-designed, will make the guitar more durable, but might affect the sound. Every product has a design lifetime. For example, cars in the US are often tested to 150,000 miles (241,400 km) since that's the assumed working life. With careful maintenance, some last longer. The design lifetime of airplanes is measured in flight hours—the F-15E is rated for 16,000 flight hours.

Few guitarists not named Willie Nelson expect a guitar to last for the working life of its player under hard use. Designing a guitar to be easy to maintain will extend its working life. For example, bolt-on necks are much easier to reset than glued necks. Another option is to glue the top on with hot hide glue, which is reversible. This makes it much easier to remove and repair a damaged or deformed top.

6.5.3 *Cracks*

Wood expands when it absorbs water from humid air and contracts as it gives up moisture to dry air. A thin plate by itself is free to expand and contract as it needs to. Internal stresses can cause bowing, warping, or even cracking, but wood selected for musical instruments is generally free of these defects.

However, when that same plate is part of a guitar, it is not as free to change dimensions and this can create significant internal stresses. If a wood guitar gets too dry, wood can crack along grain lines. Cracked guitars have been a challenge for luthiers from the beginning and are still a steady source of income for guitar repair people.

Wood shrinks more in the tangential than in the radial direction and hardly at all in the longitudinal direction (Fig. 6.124). For a quarter sawn guitar top, this means that wood expands and contracts the most laterally across to the top. This, combined with the fact that wood is weaker across the grain than along it, means that acoustic guitars sometimes crack. Mahogany (Swietenia macrophylla) has the smallest tangential to radial shrinkage ratio of commonly used woods.

Figure 6.125 shows a typical crack in the top of an acoustic guitar. As with most cracks, it was caused by the owner keeping the guitar in very dry conditions. In this case, the weather was very cold, so the air was unable to carry very much water. When that dry air was warmed up by the heater, its carrying capacity increased, but the water content didn't increase and the relative humidity was lowered. The relative humidity in the apartment where the guitar was stored was probably less than 20%, too low for a wood guitar.

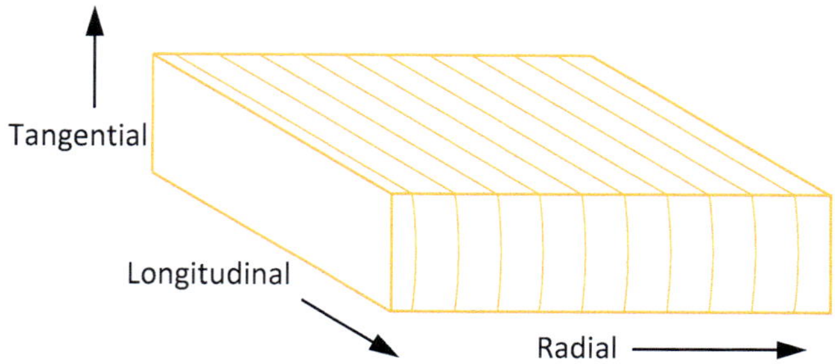

Fig. 6.124 Directions of wood grain in a board

Fig. 6.125 A crack caused by very low humidity

Contraction of the top wasn't matched by contraction in the rest of the structure and internal stresses built up. The top is braced with the same species of spruce used for the top, but the grain of the braces runs in different directions than the grain of the top. Also, the top is glued to maple sides whose contraction didn't match the top. Eventually, the internal stresses were higher than the cross-grain failure stress of the spruce top and it cracked.

Another interesting failure in this guitar was that the binding separated from the back in the lower bout (Fig. 6.126). The wood binding couldn't contract to match the contraction of the back and the glue joint failed.

Curved plates are less stiff laterally than flat plates, so one might guess that archtop guitars are slightly less susceptible to cracking than flattops. Bob Bendetto observes:

> During my days on the bench I have seen and repaired many cracked archtops … including some of my own. I especially enjoyed repairing and restoring old violins and other bowed instruments. Some a few hundred years old with repairs on top of repairs. Lots of cracks! Carved tops and backs are marginally less likely to crack than flat tops, but they do crack.

Fig. 6.126 Wood binding sprung loose by low humidity

Designing against cracking is one of the ongoing challenges facing luthiers. Three possible approaches are:

- Thermally treating wood to reduce sensitivity to changes in humidity.
- Laminating tops so that one or more cross grain plies prevent cracking.
- Replacing wood tops with fiber composites like graphite.

Thermal treatment, variously called baking, roasting, or torrefaction, changes the micromechanical structure of the wood. Builders sometimes state that torrefied tops sound like ones that have naturally aged. The basic process dates to the early nineteenth century and was originally used to treat biomass to make it a better fuel. It chemically changes the wood, increasing the weight fractions of lignin and decreasing the weight fractions of the different types of cellulose. Wood treated this way has lower density, lower elastic modulus, and lower strength. It also is generally assumed to reduce damping, though data in the technical literature is limited. Finally, it also darkens the wood. At this writing, thermally treated wood is increasingly being offered as an option on acoustic guitars and some suppliers now offer torrefied top plates.

Bob Taylor has experimented with methods to reduce cracking, including thermal treatment and laminated tops. He says:

> Glue layers are a sound killer. Perhaps hide glue would be better. But I find more success at eliminating cracks by using solid wood and baking the wood. In fact, cycling it a couple times improves it even more.

Laminated tops are most often used on the most inexpensive acoustic guitars and builders widely prefer solid wood tops. Laminated wood tops have different stiffness properties because of the cross plies, though this may be addressed by modifying the bracing pattern. A more difficult problem is glue layers that might add unwanted damping. Bob Benedetto is one of the few top builders who make instruments with laminated tops. He states:

> Plywood tops can produce a reasonably good sound if acoustics are taken into consideration when laminating. I have made many laminated tops and backs and always got pretty good results with the use of spruce veneer rather than maple or any other hard wood for the top. Fewer laminates on tops and backs (like three rather than five) will result in a thinner, more flexible and more acoustically responsive guitar. The top would also have to be braced accordingly. If one wants to make an acoustically responsive laminated guitar, then they have to abandon typical plywood construction (strength and durability) and substitute it with a balance of structure and acoustic design… the same mindset as when making carved tops and backs.
>
> My Bravo models (Fig. 6.127), with their carefully laminated construction, are a good examples of archtop guitars with laminated tops and backs that sound pretty good acoustically.

Fig. 6.127 A laminated top for a Benedetto Bravo (image courtesy of Bob Benedetto, benedettoguitars.com)

Fig. 6.128 Heavily worn frets (image courtesy of Luis Munoz, modernguitartech.com)

6.5.4 Worn Frets

Frets wear with extended use and eventually need to be retouched or replaced. Figure 6.128 shows heavily worn frets that need replacing. A228 music wire used for steel guitar strings is very strong, but hard, so it wears down frets over time. Nylon strings are much softer than fret wire, so fret wear is a less of a problem on classical guitars.

Fret wire is available in several different compositions. The most common is colloquially called nickel silver, though it has no actual silver in it. Rather, it is mostly nickel and copper along with small amounts of other metals. For example, 18% nickel silver fret wire is often 18% nickel and 80% copper. Some formulations include zinc as well.

Fret wire needs to be hard enough to withstand contact of steel guitar strings and be corrosion resistant while being flexible enough to bend easily to the needed radius and soft enough to work with steel files. Nickel silver fret wire meets these disparate requirements and holds up well under normal use. Guitars may need to be refretted only a few times in their working lifetimes. Of course, heavy playing shortens the life of the frets. If there is enough material left, frets can be filed down (redressed) to remove grooves before needing to be replaced.

One solution is to use harder fret wire. Harder wire may increase the wear on strings, but it's easier to change strings than frets. Most suppliers offer stainless steel fret wire, composed mostly of iron, chromium, and nickel, that is significantly harder than nickel silver. It is more difficult to work with and diamond files may be needed.

An alternative is wire made from CuSnFe1Ti, which is copper, tin, iron, and titanium. A popular version is EVO wire by Jescar. This wire is easy to distinguish because it is gold colored. It is harder than nickel silver, but softer than stainless steel. It is easier to work than stainless steel.

Dressing frets by filing them down to eliminate grooves necessarily removes material. An easy way to extend the life of frets is to make them taller so there is

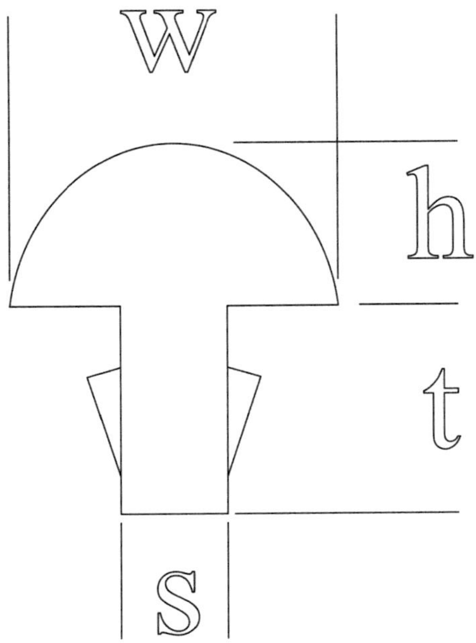

Fig. 6.129 Fret wire dimensions

Table 6.6 Representative guitar fret wire dimensions

Type	Unit	Crown width, *w*	Crown height, *h*	Tang height, *t*	Tang width, *s*
StewMac Medium/Medium	in.	0.084	0.039	0.055	0.023[a]
	mm	2.13	0.99	1.40	0.58
StewMac Medium/High	in.	0.095	0.045	0.073	0.023[a]
	mm	2.41	1.14	1.85	0.58
StewMac Medium/Higher	in.	0.092	0.048	0.062	0.023[a]
	mm	2.34	1.22	1.57	0.58
Jescar EVO 37080-EVO	in.	0.080	0.037	0.058	0.019
	mm	2.03	0.94	1.47	0.48
Luthiers Mercantile FW74	in.	0.080	0.043	0.050	0.020
	mm	2.03	1.09	1.27	0.51

[a]Stewart MacDonald fret wire is sized for 0.023 in. wide fret slots. Tang is slightly narrower

more material to remove when redressing. However, this changes the feel of the instrument and some players may not like the higher frets.

Fret wire is available in an almost bewildering number of sizes. However, it can be reduced to a few dimensions, as shown in Fig. 6.129.

Representative dimensions are shown in Table 6.6.

Bibliography

Books

Benedetto, R., Making an Archtop Guitar, 2nd ed., Centerstream Publishing , 2018, ISBN 978-1-57424-355-0

Forest Products Lab, (n.d.), Wood Handbook: Wood as an Engineering Material, General Technical Report FPL, GTR 113, US Dept of Agriculture, Forest Service, Forest Products Laboratory. Fs.usda.gov/treesearch/pubs/5734

Gore, T. and Gilet, G., Contemporary Acoustic Guitar Design and Build, 2nd ed., published by Trevor Gore, n.d., ISBN Vol 1: 978-0-9871174-2-7, Vol 2: 978-0-9871174-3-4

Marra, A.A., Technology of Wood Bonding: Principles in Practice, Van Nostrand Reinhold, New York, 1992.

Westbrook, J., Guitar Making in Nineteenth-Century London: Louis Panormo and his Contemporaries (PhD dissertation, The University of Cambridge, 2012). Available at: https://www.repository.cam.ac.uk/handle/1810/283918

Articles

Bartolini, W. and Bartolini, P.A., Experimental Studies of Classic and Flamenco Guitars, Journal of Guitar Acoustics, No. 6, September 1982, pp. 74-103.

Blilie, Jim, Proposed Standard Glue Strength Testing Method, American Lutherie, n.d., #126, Summer 2016, pp38-47

Bourgois, J., Bartholin, M.C. and Guyonnet, R., Thermal Treatment of Wood: Analysis of the Obtained Product, Wood Science and Technology, Vol. 23, 1989, pp. 303-310

Brémaud, I., Acoustical Properties of Wood in String Instruments Soundboards and Tuned Idiophones: Biological and Cultural Diversity, Journal of the Acoustical Society, Vol. 131, No. 807, 2012, doi: https://doi.org/10.1121/1.3651233

Brown, L.D., White, Yellow and Hide Glues, Guild of American Luthiers Data Sheet #174, also Big Red Book of American Lutherie, n.d., Vol. 1, ISBN 0-9626447-2-2

Brunė, R.E., Eight Concerns of Highly Successful Guitar Makers, American Lutherie #79, 2004, Also Big Red Book of American Lutherie, Vol. 7, ISBN 978-0-9968015-0-8

Byers, G., Classical Guitar Intonation, American Lutherie, #47, 1006, p34, also Big Red Book of American Lutherie, n.d., Vol. 4, ISBN 978-0-9626447-6-4

Carcagno, S., Bucknall, R., Woodhouse, J., Fritz, C. and Plack, C.J., Effect of Back Wood Choice on the Perceived Quality of Steel-String Acoustic Guitars, The Journal of the Acoustical Society of America, Vol. 144, No. 3533, 2018, doi: https://doi.org/10.1121/1.5084735

Darnton, Michael, The Power of Circles, American Lutherie, #87, Fall 2006, p26

Doolin, M., Intonation in the Real World, American Lutherie, #92, 2007, pp. 26-34

Durfee, D.S. and Colton, J.S., The Physics of Musical Scales: Theory and Experiment, American Journal of Physics, Vol. 83, No. 835, 2015, doi: https://doi.org/10.1119/1.4926956

Efendy, E. and French, M., Reducing Build Variation in Arched Plates, Experimental and Applied Mechanics, Vol. 6, Conference proceedings of the Society for Experimental Mechanics, n.d., pp. 607–620, doi: https://doi.org/10.1007/978-1-4419-9792-0_89

Esteves, B.M. and Pereira, H.M., Wood Modification by Heat Treatment: A Review, BioResources, n.d., Vol. 4, No. 1, pp 370–404

Firth, I.M., Physics of the Guitar at the Helmholtz and First Top Plate Resonance, The Journal of the Acoustical Society of America, Vol. 61, No. 588, 1977, doi: https://doi.org/10.1121/1.381302

Ford, F. and MacRostie, D., Stick with Hide Glue, American Lutherie, #57, 1999, also Big Red Book of American Lutherie, Vol. 5, ISBN 978-0-9626447-7-1
French, M., A Better Approximation to the Rule of 18, American Lutherie, #132, n.d., p. 62
French, M. and Brubaker, K., Build Variation in a Group of Acoustic Guitars, American Lutherie, #90, p28, also Flattop Guitars, Guild of American Luthiers, p72
French, M., An Improved Body Shape Definition for Acoustic Guitars, Savart Journal, Vol. 1, No. 6, 2017, https://www.savartjournal.org/index.php/sj/article/view/26, also https://arxiv.org/abs/1802.00439
Frihart, C.R. and Beecher, J.F., Factors That Lead to Failure with Wood Adhesive Bonds, Proceedings of 2016 World Conference on Timber Engineering, Vienna
Gilbert, J., Intonation and Fret Placement, Soundboard, Spring 1984, pp. 26-27
Gore, T., Wood for Guitars, Proceedings of the 161st Meeting of the Acoustical Society of America, Seattle, May 2011, doi: https://doi.org/10.1121/1.3610500
Hunt, C.G, Frihart, C.R., Dunky, M. Rohumaa, A., Understanding Wood Bonds – Going Beyond What Meets the Eye: A Critical Review, Review of Adhesion and Adhesives, Vol. 6 , No. 4, 2018, pp. 369-440, DOI: https://doi.org/10.7569/RAA.2018.097312
Jacobson, P., Epoxy, Guild of American Luthiers Data Sheet #90, also Big Red Book of American Lutherie, n.d., Vol. 1, ISBN 0-9626447-2-2.
Magliari, G. and MacRostie, D., Beyond the Rule of 18: Intonation for the 21st Century, American Lutherie, #116, 2013, p. 6
Meeker, Ward, Mark Erlewine's 40 Years with Willie Nelson's Trigger, Vintage Guitar Magazine,
Merchel, S., Altinsoy, M.E. and Olson, D., Perceptual Evaluation of Bracewood and Soundboard Wood Variations on the Preference of a Steel-String Acoustic Guitar, The Journal of the Acoustical Society of America, Vol. 146, No. 2608, 2019, doi: https://doi.org/10.1121/1.5129395
Mottola, R.M., Same-Fretted-Note Intonation Variability of the Steel String Acoustic Guitar, n.d., Retrieved from OSF Preprint Server 1/6/2021, https://osf.io/45mtc/
Musser, D., Taking the Guitar Beyond Equal Temperament, American Lutherie #30, 1992, Also Big Red Book of American Lutherie, Vol. 3, ISBN 0-9656447-5-7.
Polievkt, P., Johnson, W. and Perova-Mello, N., The Physics of Guitar String Vibrations, The American Journal of Physics, Vol. 84, No. 38, 2016, https://doi.org/10.1119/1.4935088
Torres, J.A. and Boullosa, R.R., Influence of the Bridge on the Vibrations of the Top Plate of a Classical Guitar, Applied Acoustics, Vol. 70, Issues 11-12, December 2009, pp 1371-1377, doi: https://doi.org/10.1016/j.apacoust.2009.07.002
Varieschi, G.U. and Gower, C.M., Intonation and Compensation of Fretted String Instruments, American Journal of Physics, Vol. 78, No. 47, 2010, doi: https://doi.org/10.1119/1.3226563
Viala, R., Placet, V. and Cogan, S., Simultaneous Non-destructive Identification of Multiple Elastic and Damping Properties of Spruce Tonewood to Improve Grading, Journal of Cultural Heritage, Vol. 42, 2020, pp. 108-116
Wegst, U.G.K., Wood for Sound, American Journal of Botany, Vol. 92, No. 10, 2006, pp. 1439-1448, doi: https://doi.org/10.3732/ajb.93.10.1439
Yarosh, B., Adjustable Saddles for Acoustic Guitars, American Lutherie #82, 2005, Also Big Red Book of American Lutherie, Vol. 7, ISBN 978-0-9968015-0-8
Yoshikawa, S., Acoustical Classification of Wood for Stringed Instruments, Journal of the Acoustical Society of America, Vol. 122, No. 568, 2007, https://doi.org/10.1121/1.2743162
Yu Q., Mostaghel, N. and Fu, K-C, Effect of Initial Curvature on Natural Frequency of Thin Plate on Hinge Supports, Journal of Engineering Mechanics, Vol. 120, No. 4, 1994, pp. 796-813

Chapter 7
Technical Reference Information

This section lists useful reference information. Some of it appears elsewhere, but is repeated here for convenience.

R. M. French, *Acoustic Guitar Design*,
https://doi.org/10.1007/978-3-030-89381-1_7

7.1 Frequencies of Notes in the Human Hearing Range (Hz)

E_0	20.602	E_3	164.81	E_6	1318.5	E_9	10548
F_0	21.827	F_3	174.61	F_6	1396.9	F_9	11175
	23.125		185.00		1480.0		11840
G_0	24.500	G_3	196.00	G_6	1568.0	G_9	12544
	25.957		207.65		1661.2		13290
A_0	27.500	A_3	220.00	A_6	1760.0	A_9	14080
	29.135		233.08		1864.7		14917
B_0	30.868	B_3	246.94	B_6	1975.5	B_9	15804
C_1	32.703	C_4	261.63	C_7	2093.0	C_{10}	16744
	34.648		277.18		2217.5		17740
D_1	36.708	D_4	293.66	D_7	2349.3	D_{10}	18795
	38.891		311.13		2489.0		19912
E_1	41.203	E_4	329.63	E_7	2637.0	E_{10}	21096
F_1	43.654	F_4	349.23	F_7	2793.8		
	46.249		369.99		2960.0		
G_1	48.999	G_4	392.00	G_7	3136.0		
	51.913		415.30		3322.4		
A_1	55.000	A_4	440.00	A_7	3520.0		
	58.270		466.16		3729.3		
B_1	61.735	B_4	493.88	B_7	3951.1		
C_2	65.406	C_5	523.25	C_8	4186.0		
	69.296		554.37		4434.9		
D_2	73.416	D_5	587.33	D_8	4698.6		
	77.782		622.25		4978.0		
E_2	82.407	E_5	659.26	E_8	5274.0		
F_2	87.307	F_5	698.46	F_8	5587.7		
	92.499		739.99		5919.9		
G_2	97.999	G_5	783.99	G_8	6271.9		
	103.83		830.61		6644.9		
A_2	110.00	A_5	880.00	A_8	7040.0		
	116.54		932.33		7458.6		
B_2	123.47	B_5	987.8	B_8	7902.1		
C_3	130.81	C_6	1046.5	C_9	8372.0		
	138.59		1108.7		8869.8		
D_3	146.83	D_6	1174.7	D_9	9397.3		
	155.56		1244.5		9956.1		

7.2 Fret Locations

Fret positions are measured from the nut, since the saddle is generally intonated. The expression for fret location is

$$L_n = L_0\left(1 - \frac{1}{r^n}\right) \quad \text{where } r = \sqrt[12]{2} \approx 1.05946$$

L_0 is the scale length and n is the fret number. This expression works in any units, so it can be applied for inches and millimeters. Table 7.1 shows fret locations for several common scale lengths. To calculate locations for other scale lengths, simply multiply the scale length by the coefficients in the second column.

Table 7.1 Fret locations for some common scale lengths

			in.				mm	
Fret	Coefficient	24	24.75	25.4	25.5	630	650	660
1	0.0561	1.347	1.389	1.426	1.431	35.359	36.482	37.043
2	0.1091	2.618	2.700	2.771	2.782	68.734	70.916	72.007
3	0.1591	3.818	3.938	4.041	4.057	100.235	103.417	105.008
4	0.2063	4.951	5.106	5.240	5.261	129.969	134.095	136.158
5	0.2508	6.020	6.208	6.372	6.397	158.033	163.050	165.559
6	0.2929	7.029	7.249	7.439	7.469	184.523	190.381	193.310
7	0.3326	7.982	8.231	8.448	8.481	209.525	216.177	219.503
8	0.3700	8.881	9.158	9.399	9.436	233.125	240.526	244.226
9	0.4054	9.730	10.034	10.297	10.338	255.400	263.508	267.562
10	0.4388	10.530	10.860	11.145	11.189	276.424	285.200	289.588
11	0.4703	11.286	11.639	11.945	11.992	296.269	305.674	310.377
12	0.5000	12.000	12.375	12.700	12.750	315.000	325.000	330.000
13	0.5281	12.674	13.070	13.413	13.466	332.680	343.241	348.521
14	0.5546	13.309	13.725	14.086	14.141	349.367	360.458	366.003
15	0.5796	13.909	14.344	14.721	14.779	365.118	376.709	382.504
16	0.6031	14.476	14.928	15.320	15.380	379.984	392.047	398.079
17	0.6254	15.010	15.479	15.886	15.948	394.017	406.525	412.779
18	0.6464	15.515	16.000	16.420	16.484	407.261	420.190	426.655
19	0.6663	15.991	16.491	16.924	16.990	419.763	433.089	439.751
20	0.6850	16.440	16.954	17.400	17.468	431.562	445.263	452.113
21	0.7027	16.865	17.392	17.849	17.919	442.700	456.754	463.781
22	0.7194	17.265	17.805	18.272	18.344	453.212	467.600	474.794
23	0.7351	17.643	18.195	18.672	18.746	463.135	477.837	485.189
24	0.7500	18.000	18.563	19.050	19.125	472.500	487.500	495.000

7.3 Classical Guitars

The following dimensions are representative of full-size classical guitars and are intended as a reference for designers. It's unlikely that any single instrument would have all of these dimensions exactly. Rather, a designer can check these values to make sure they are on the right track (Tables 7.2, 7.3, and 7.4).

Table 7.2 Typical dimensions for classical guitars

Dimension	Value (in.)	Value (mm)
Scale length	25.2–25.98	640–660
Headstock width—top	2.87	73
Headstock width—bottom	2.36	60
Headstock thickness (including head plate)	0.79	20
Fretboard thickness	0.24	6.0
Neck thickness at nut (excluding fretboard)	0.63	16
Neck thickness at eighth fret (excluding fretboard)	0.768	19.5
Neck width at 12th fret	2.40	61
Body length	19.25	489
Soundhole diameter	3.39	86
Soundhole center location from nut	18.7	475
Soundhole center location from 12th fret	5.93	151
Body depth at tail block	3.86	98
Body depth at neck block	3.46	88
Upper bout width	11.02	280
Width at waist	9.25	235
Lower bout width	14.37	365
Side thickness	0.079	2.0
Back thickness	0.094	2.4
Top thickness	0.094–0.126	2.4–3.2
Nut width	1.97–2.13	50–54
Bridge height	0.35	9
String spacing at nut (outside centers)	1.69–1.93	43–49
String spacing at saddle (outside centers)	2.24–2.36	57–60
Tuner roller diameter (nom)	0.394	10
Tuner roller spacing (nom)	1.375	35

Table 7.3 Nylon string diameters

Type	Unit	E	A	D	G	B	E
D'Addario Pro Arté	in.	0.042	0.033	0.028	0.0397	0.0317	0.0275
Nylon Core Light Tension	mm	1.067	0.838	0.711	1.008	0.805	0.699
D'Addario Pro Arté	in.	0.044	0.035	0.028	0.0403	0.0322	0.028
Dynacore Normal Tension	mm	1.118	0.889	0.711	1.024	0.818	0.711
D'Addario Pro Arté	in.	0.046	0.036	0.029	0.0409	0.0327	0.0285
Dynacore Hard Tension	mm	1.168	0.914	0.737	1.039	0.831	0.724
D'Addario Pro Arté	in.	0.047	0.036	0.030	0.0415	0.0333	0.029
Dynacore Extra Hard Tension	mm	1.194	0.914	0.762	1.054	0.846	0.737

Table 7.4 Approximate string action for classical guitars

Position	Unit	Bass E	Treble E
1st Fret	in.	0.030	0.024
	mm	0.762	0.610
12th Fret	in.	0.156	0.125
	mm	3.962	3.175

Neck relief of 0.002 in. (0.05 mm) at the eighth fret is typical in classical guitars.

7.4 Steel String Guitars

Again, these dimensions are intended as a reference and it's unlikely that any one steel string guitar would have all of these dimensions exactly (Tables 7.5, 7.6, and 7.7).

Neck relief of 0.002 in. (0.05 mm) at the eighth fret is typical in steel string guitars.

7.5 General Information

Note that wood properties are variable and these numbers are only representative.

Stiffness per unit weight is

$$k = \frac{EI_x}{\rho A} = \frac{E\frac{1}{12}wh^3}{\rho wh} = \frac{Eh^2}{12\rho}$$

where E is elastic modulus, I_x is area moment of inertia about the x axis, ρ is density, A is cross-sectional area, w is width, and h is height. Length, L, does not appear in this expression. The bending stiffness of beams is inversely proportional to L^3.

Table 7.5 Typical dimensions for full-sized steel string acoustic guitars (OM Body)

Dimension	Value (in.)	Value (mm)
Scale length	24–25.5	609.6–647.7
Headstock width—top	2.875 (2 7/8)	73
Headstock width—bottom	2.25 (2¼)	57.2
Headstock thickness (including head plate)	0.625 (5/8)	15.9
Fretboard Thickness	0.25 (1/4)	6.35
Neck thickness at first fret (excluding fretboard)	0.64	16.3
Neck thickness at eighth fret (excluding fretboard)	¾ (0.75)	19.1
Neck width at 12th fret	2.2	55.9
Body length	19.25	489
Soundhole diameter	3.875 (3 7/8)	98.4
Soundhole center location from nut	19.875 (19 7/8)	504.8
Soundhole center location from 14th fret	5.90	150
Body Depth at tail block	4.125 (4 1/8)	104.8
Body Depth at neck block	3.25 (3 ¼)	82.6
Upper bout width	11.25	285.8
Width at waist	9 1/4 (9.25)	235
Lower bout width	15	381
Side thickness	0.085–0.105	2.2–2.7
Back thickness	0.095–0.105	2.4–2.7
Top thickness	0.105–0.120	2.7–3.0
Nut width	1.69–1.75	42.9–44.5
Bridge height	3/8 (0.375)	9.5
String spacing at nut (outside centers)	1 7/16 (1.438)	36.5
String spacing at saddle (outside centers)	2.125–2.313	54.0–58.8
Tuner spacing	1.625 (1 5/8)	41.3
Top radius	25 ft. – 40 ft	7.62 m – 12.2 m
Back radius	12 ft. – 25 ft	3.66 m – 7.62 m

Table 7.6 Steel string diameters

Type	Unit	E	A	D	G	B	E
D'Addario Phosphor Bronze	in.	0.047	0.039	0.030	0.023	0.014	0.010
Extra Light Gauge	mm	1.194	0.991	0.762	0.584	0.356	0.254
D'Addario Phosphor Bronze	in.	0.053	0.042	0.032	0.024	0.016	0.012
Light Gauge	mm	1.346	1.067	0.813	0.610	0.406	0.305
D'Addario Phosphor Bronze	in.	0.056	0.045	0.035	0.026	0.017	0.013
Medium Gauge	mm	1.422	1.143	0.889	0.660	0.432	0.330
D'Addario Phosphor Bronze	in.	0.059	0.049	0.039	0.027	0.018	0.014
Heavy Gauge	mm	1.499	1.245	0.991	0.686	0.457	0.356

The normalized stiffness of rectangular bar stock is EI/ρA.

Table 7.7 Approximate string action for steel string guitars

Position	Unit	Bass E	Treble E
1st Fret	in.	0.023	0.013
	mm	0.584	0.330
12th Fret	in.	0.100	0.079
	mm	2.54	2.0

Table 7.9 Stiffness per unit weight of rectangular bar stock

Width, w (mm)	Height, h (mm)	6061-T6 Aluminum	Carbon Fiber	1010 Mild Steel	Titanium	Honduran Mahogany
2	2	8.51	28.22	8.70	8.59	6.81
2	4	34.02	112.89	34.82	34.87	27.26
3	3	19.14	63.5	19.59	19.33	15.33
3	6	76.56	254.0	78.34	77.33	61.33
4	4	34.02	112.89	34.82	34.37	27.26
4	8	136.1	451.56	139.28	137.48	109.04

Table 7.8 Mechanical properties of common materials

	Density		Modulus	
Species	g/cm^3	lb/in.^{3}	Gpa	ksi
Sitka Spruce	0.36 (12%)	0.013	9.9	1440
Western Red Cedar	0.32 (12%)	0.012	7.7	1120
Honduran Mahogany	0.6 (12%)	0.022	10.3 (Flexural)	1490
Unidirectional Carbon Fiber	1.7	0.061	127	18,500
1010 Mild Steel	7.87	0.284	205	29,700
6061-T6 Aluminum	2.70	0.098	68.9	10,000
Titanium Grade 2	4.51	0.163	103	14,900

Correction to: Detail Design

Correction to:
Chapter 6 in: R. M. French, *Acoustic Guitar Design*,
https://doi.org/10.1007/978-3-030-89381-1_6

An incorrect character was found in the chapter and it has been updated now.

The updated version of this chapter can be found at
https://doi.org/10.1007/978-3-030-89381-1_6

R. M. French, *Acoustic Guitar Design*,
https://doi.org/10.1007/978-3-030-89381-1_8

Index

R. M. French, *Acoustic Guitar Design*,
https://doi.org/10.1007/978-3-030-89381-1